™

P9-BYV-923

References for the Rest of Us™

BUSINESS BOOK SERIES FROM IDG

Are you intimidated and confused by personal finance and business issues? Do you find that traditional books are overloaded with technical details and advice that you'll never use? Do you postpone important financial decisions because you just don't want to deal with them? Then our ...*For Dummies*™ business book series is for you.

...*For Dummies*™ business books are written for those frustrated and hard-working souls who know they aren't really dumb but find that the myriad of personal finance and business issues and the accompanying horror stories make them feel helpless. ...*For Dummies*™ books use a lighthearted approach, a down-to-earth style, and even cartoons and humorous icons to diffuse fears and build confidence. Lighthearted but not lightweight, these books are perfect survival guides to solve your personal finance and business problems.

> *"More than a publishing phenomenon, 'Dummies' is a sign of the times."*
> — The New York Times

> *"...you won't go wrong buying them."*
> — Walter Mossberg, Wall Street Journal, on IDG's ...For Dummies™ books

> *"This is the best book ever written for a beginner."*
> — Clarence Petersen, Chicago Tribune, on DOS For Dummies ®

Already, hundreds of thousands of satisfied readers agree. They have made ...*For Dummies*™ the #1 introductory level computer book series and a best-selling business book series. They have written asking for more. So if you're looking for the best and easiest way to learn about personal finance and business, look to ...*For Dummies*™ to give you a helping hand.

™

IDG BOOKS

EVERYDAY
MATH
FOR
DUMMIES™

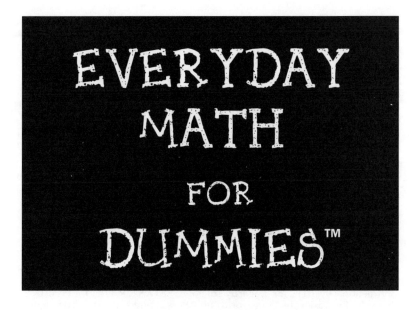

EVERYDAY MATH FOR DUMMIES™

by Charles Seiter

IDG Books Worldwide, Inc.
An International Data Group Company

Foster City, CA ♦ Chicago, IL ♦ Indianapolis, IN ♦ Braintree, MA ♦ Dallas, TX

Everyday Math For Dummies™

Published by
IDG Books Worldwide, Inc.
An International Data Group Company
919 East Hillsdale Blvd.
Suite 400
Foster City, CA 94404

Library of Congress Catalog Card No.: 95-77881

ISBN: 1-56884-248-1

Printed in the United States of America

10 9 8 7 6 5 4 3 2 1

1E/RT/QU/ZV

Distributed in the United States by IDG Books Worldwide, Inc.

Distributed by Macmillan Canada for Canada; by Computer and Technical Books for the Caribbean Basin; by Contemporantea de Ediciones for Venezuela; by Distribuidora Cuspide for Argentina; by CITFC for Brazil; by Ediciones ZETA S.C.R. Ltda. for Peru; by Editorial Limusa SA for Mexico; by Transworld Publishers Limited in the United Kingdom and Europe; by Al-Maiman Publishers & Distributors for Saudi Arabia; by Simron Pty. Ltd. for South Africa; by IDG Communications (HK) Ltd. for Hong Kong; by Toppan Company Ltd. for Japan; by Addison Wesley Publishing Company for Korea; by Longman Singapore Publisher Ltd. for Singapore, Malaysia, Thailand, and Indonesia; by Unalis Corporation for Taiwan; by WS Computer Publishing Company, Inc. for the Philippines; by WoodsLane Enterprises Ltd. for New Zealand.

For general information on IDG Books in the U.S., including information on discounts and premiums, contact IDG Books at 800-434-3422 or 415-655-3000.

For information on where to purchase IDG Books outside the U.S., contact IDG Books International at 415-655-3021 or fax 415-655-3295.

For information on translations, contact Marc Jeffrey Mikulich, Director, Foreign and Subsidiary Rights, at IDG Books Worldwide, 415-655-3018 or fax 415-655-3295.

For sales inquiries and special prices for bulk quantities, write to the address above or call IDG Books Worldwide at 415-655-3000.

For information on using IDG Books in the classroom, or for ordering examination copies, contact Jim Kelly at 800-434-2086.

 ™ is a trademark under exclusive license to IDG Books Worldwide, Inc., from International Data Group, Inc.

About the Author

Charles Seiter

Charles Seiter wrote his first math book at the age of ten. It was a cartoon guide to calculus that was used as part of a junior-college math course. Since then, he has written 14 books (most of them are computer books), with two more on the way in 1995.

Along the way, he got a Ph.D. from Caltech but later abandoned academic life, with the help of TV game show money, and spent several years designing and teaching non-standard adult math classes in business math, statistics, and calculus. His message is simple: *Anyone* can learn everyday math with the right approach. Even if you had your math confidence surgically removed in high school, *Everyday Math For Dummies* can replace it!

ABOUT
IDG
BOOKS
WORLDWIDE

Welcome to the world of IDG Books Worldwide.

IDG Books Worldwide, Inc., is a subsidiary of International Data Group, the world's largest publisher of computer-related information and the leading global provider of information services on information technology. IDG was founded more than 25 years ago and now employs more than 7,200 people worldwide. IDG publishes more than 233 computer publications in 65 countries (see listing below). More than sixty million people read one or more IDG publications each month.

Launched in 1990, IDG Books Worldwide is today the #1 publisher of best-selling computer books in the United States. We are proud to have received 3 awards from the Computer Press Association in recognition of editorial excellence, and our best-selling ...For Dummies™ series has more than 12 million copies in print with translations in 25 languages. IDG Books, through a recent joint venture with IDG's Hi-Tech Beijing, became the first U.S. publisher to publish a computer book in the People's Republic of China. In record time, IDG Books has become the first choice for millions of readers around the world who want to learn how to better manage their businesses.

Our mission is simple: Every IDG book is designed to bring extra value and skill-building instructions to the reader. Our books are written by experts who understand and care about our readers. The knowledge base of our editorial staff comes from years of experience in publishing, education, and journalism — experience which we use to produce books for the '90s. In short, we care about books, so we attract the best people. We devote special attention to details such as audience, interior design, use of icons, and illustrations. And because we use an efficient process of authoring, editing, and desktop publishing our books electronically, we can spend more time ensuring superior content and spend less time on the technicalities of making books.

You can count on our commitment to deliver high-quality books at competitive prices on topics consumers want to read about. At IDG, we value quality, and we have been delivering quality for more than 25 years. You'll find no better book on a subject than an IDG book.

John Kilcullen
President and CEO
IDG Books Worldwide, Inc.

WINNER
Eighth Annual
Computer Press
Awards ≥ 1992

WINNER
Ninth Annual
Computer Press
Awards ≥ 1993

IDG
BOOKS

IDG Books Worldwide, Inc., is a subsidiary of International Data Group, the world's largest publisher of computer-related information and the leading global provider of information services on information technology. International Data Group publishes over 220 computer publications in 65 countries. More than fifty million people read one or more International Data Group publications each month. The officers are Patrick J. McGovern, Founder and Board Chairman; Kelly Conlin, President; Jim Casella, Chief Operating Officer. International Data Group's publications include: **ARGENTINA'S** Computerworld Argentina, Infoworld Argentina; **AUSTRALIA'S** Computerworld Australia, Computer Living, Australian PC World, Australian Macworld, Network World, Mobile Business Australia, Publish!, Reseller, IDG Sources; **AUSTRIA'S** Computerwelt Oesterreich, PC Test; **BELGIUM'S** Data News (CW); **BOLIVIA'S** Computerworld; **BRAZIL'S** Computerworld, Connections, Game Power, Mundo Unix, PC World, Publish, Super Game; **BULGARIA'S** Computerworld Bulgaria, PC & Mac World Bulgaria, Network World Bulgaria; **CANADA'S** CIO Canada, Computerworld Canada, InfoCanada, Network World Canada, Reseller; **CHILE'S** Computerworld Chile, Informatica; **COLOMBIA'S** Computerworld Colombia, PC World; **COSTA RICA'S** PC World; **CZECH REPUBLIC'S** Computerworld, Elektronika, PC World; **DENMARK'S** Communications World, Computerworld Danmark, Computerworld Focus, Macintosh Produktkatalog, Macworld Danmark, PC World Danmark, PC Produktguide, Tech World, Windows World; **ECUADOR'S** PC World Ecuador; **EGYPT'S** Computerworld (CW) Middle East, PC World Middle East; **FINLAND'S** MikroPC, Tietoviikko, Tietoverkko; **FRANCE'S** Distributique, GOLDEN MAC, InfoPC, Le Guide du Monde Informatique, Le Monde Informatique, Telecoms & Reseaux; **GERMANY'S** Computerwoche, Computerwoche Focus, Computerwoche Extra, Electronic Entertainment, Gamepro, Information Management, Macwelt, Netzwelt, PC Welt, Publish, Publish; **GREECE'S** Publish & Macworld; **HONG KONG'S** Computerworld Hong Kong, PC World Hong Kong; **HUNGARY'S** Computerworld SZT, PC World; **INDIA'S** Computers & Communications; **INDONESIA'S** Info Komputer; **IRELAND'S** ComputerScope; **ISRAEL'S** Beyond Windows, Computerworld Israel, Multimedia, PC World Israel; **ITALY'S** Computerworld Italia, Lotus Magazine, Macworld Italia, Networking Italia, PC Shopping Italy, PC World Italia; **JAPAN'S** Computerworld Today, Information Systems World, Macworld Japan, Nikkei Personal Computing, SunWorld Japan, Windows World; **KENYA'S** East African Computer News; **KOREA'S** Computerworld Korea, Macworld Korea, PC World Korea; **LATIN AMERICA'S** GamePro; **MALAYSIA'S** Computerworld Malaysia, PC World Malaysia; **MEXICO'S** Compu Edicion, Compu Manufactura, Computacion/Punto de Venta, Computerworld Mexico, MacWorld, Mundo Unix, PC World, Windows; **THE NETHERLANDS'** Computer! Totaal, Computable (CW), LAN Magazine, Lotus Magazine, MacWorld; **NEW ZEALAND'S** Computer Buyer, Computerworld New Zealand, Network World, New Zealand PC World; **NIGERIA'S** PC World Africa; **NORWAY'S** Computerworld Norge, Lotusworld Norge, Macworld Norge, Maxi Data, Networld, PC World Ekspress, PC World Nettverk, PC World Norge, PC World's Produktguide, Publish& Multimedia World, Student Data, Unix World, Windowsworld; **PAKISTAN'S** PC World Pakistan; **PANAMA'S** PC World Panama; **PERU'S** Computerworld Peru, PC World; **PEOPLE'S REPUBLIC OF CHINA'S** China Computerworld, China Infoworld, China PC Info Magazine, Computer Fan, PC World China, Electronics International, Electronics Today/Multimedia World, Electronic Product World, China Network World, Software World Magazine, Telecom Product World; **PHILIPPINES'** Computerworld Philippines, PC Digest (PCW); **POLAND'S** Computerworld Poland, Computerworld Special Report, Networld, PC World/Komputer, Sunworld; **PORTUGAL'S** Cerebro/PC World, Correio Informatico/Computerworld, MacIn; **ROMANIA'S** Computerworld, PC World, Telecom Romania; **RUSSIA'S** Computerworld-Moscow, Mir - PK (PCW), Sety (Networks); **SINGAPORE'S** Computerworld Southeast Asia, PC World Singapore; **SLOVENIA'S** Monitor Magazine; **SOUTH AFRICA'S** Computer Mail (CIO),Computing S.A.,Network World S.A., Software World; **SPAIN'S** Advanced Systems, Amiga World, Computerworld Espana, Communicaciones World, Macworld Espana, NeXTWORLD, Super Juegos Magazine (GamePro), PC World Espana, Publish; **SWEDEN'S** Attack, ComputerSweden, Corporate Computing, Macworld, Mikrodatorn, Natverk & Kommunikation, PC World, CAP & Design, Datalngenjoren, Maxi Data,Windows World; **SWITZERLAND'S** Computerworld Schweiz, Macworld Schweiz, PC Tip; **TAIWAN'S** Computerworld Taiwan, PC World Taiwan; **THAILAND'S** Thai Computerworld; **TURKEY'S** Computerworld Monitor, Macworld Turkiye, PC World Turkiye; **UKRAINE'S** Computerworld, Computers+Software Magazine; **UNITED KINGDOM'S** Computing /Computerworld, Connexion/Network World, Lotus Magazine, Macworld, Open Computing/Sunworld; **UNITED STATES'** Advanced Systems, AmigaWorld, Cable in the Classroom, CD Review, CIO, Computerworld, Computerworld Client/Server Journal, Digital Video, DOS World, Electronic Entertainment Magazine (E2), Federal Computer Week, Game Hits, GamePro, IDG Books, Infoworld, Laser Event, Macworld, Maximize, Multimedia World, Network World, PC Letter, PC World, Publish, SWATPro, Video Event; **URUGUAY'S** PC World Uruguay; **VENEZUELA'S** Computerworld Venezuela, PC World; **VIETNAM'S** PC World Vietnam. 02/28/95

Acknowledgments

The author would like to thank John Kilcullen, President and CEO of IDG Books Worldwide, for telling me to do this book and in fact suggesting much of the content. The parts of it you like best are probably his idea!

At IDG Books Chicago, Kathy Welton made countless useful suggestions and provided considerable schedule encouragement, and Stacy Collins made thrashing through all sorts of last minute details bearable. At IDG Books' Indianapolis office, Pam Mourouzis, ably assisted by Suzanne Packer, worked nights and weekends to get this book through various revisions in a timely manner. Thanks also go to Beth Jenkins and her Production staff.

K. Calderwood provided helpful discussions of the material in the chapter on trends. Background material on taxes was provided by Dennis Gary of Ledger Domaine in Sonoma County. L. Lewand and D. Underwood made suggestions in their own areas of expertise, tipping and gambling.

I would also suggest, based on my experiences in writing this book, that if you encounter Loretta Toth in a social situation, simply fall down and worship at her feet. It's the least we all can do.

Finally, I would like to thank math teachers everywhere, who are still waging a mostly successful battle for their students in an increasingly short-attention-span world.

(The publisher would like to thank Patrick J. McGovern and Bill Murphy, without whom this book would not have been possible.)

Credits

VP & Publisher
Kathleen A. Welton

Brand Manager
Stacy S. Collins

Editorial Assistants
Stacey Holden Prince
Kevin Spencer

Production Director
Beth Jenkins

**Suprvisor of
Project Coordination**
Cindy L. Phipps

Project Coordinator
Valery Bourke

Pre-Press Coordinator
Steve Peake

**Associate Pre-Press
Coordinator**
Tony Augsburger

Project Editor
Pamela Mourouzis

Copy Editors
Diane L. Giangrossi
Michael Kelly
Suzanne Packer

Technical Reviewer
Brian Schellenberger

Production Staff
Paul Belcastro
Sherry Gomoll
Mark Owens
Laura Puranen
Carla Radzikinas
Dwight Ramsey
Patricia R. Reynolds
Kathie S. Schnorr
Gina Scott

Proofreader
Henry Lazarek

Indexer
Steve Rath

Cover Design
Kavish + Kavish

Contents at a Glance

Cartoons at a Glance

By Rich Tennant

Page 7

Page 118

Page 220

Page 265

Page 310

More Cartoons at a Glance

By Rich Tennent

Page 81

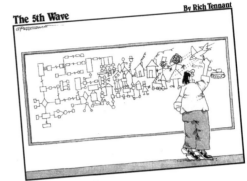

Page 135

Page 182

Page 207

Page 254

Table of Contents

· ·

Introduction

IDG Books Worldwide, Inc. has a . . . *For Dummies* title for almost every modern information need, from computer programming to personal finance. This particular title, however, is a little bit special. You could probably live a long and happy life without ever learning WordPerfect 5.1, for example, if only for the reason that WordPerfect 6.0 would have superseded it. You could even ignore my magnificent *The Internet For Macs For Dummies*. Admittedly, you would be "out of the loop" in cyberspace for the rest of your life, but you might rather learn to play the guitar than surf the Net, and I couldn't blame you a bit.

But you just can't get along in life without the material in *this* book, at least not without suffering dire consequences. If you can't balance your checkbook, you don't understand percentages, and you don't know the real odds in your state lottery, it's going to cost you, literally. You need this stuff — no two ways about it. In *Everyday Math For Dummies,* I'm trying to give you a reasonable command over 99 percent of the daily life situations that involve numbers.

Reading this book won't be like learning math in school. The emphasis here is on a relatively small number of examples, repeated in several variations so that you can find the one that's easiest for you to understand. There are no tests. I want you to get 100 percent of this material, so you won't see any trick questions, either.

Dummies — A Loaded Word

Plenty of people feel that they are "dummies" when it comes to math. Lurking in the past are all sorts of problems with math, resulting in humiliation, disappointment, scowling teachers, and dreaded exams. Many women at some time or another have been told quite pointedly that they are not expected to be good at math. A recent well-documented study of classroom practice produced the discouraging result that teachers do indeed tend to ignore girls in elementary school math classes, even if the girls are getting the best grades in the class. Female teachers are no better in this regard than male teachers, either.

The net result of this situation is math phobia or math avoidance on a massive scale. My modest ambition is simply to fix this situation with *Everyday Math For Dummies.* This book is intended to be a quick, enjoyable adult-education course in the aspects of math that you need all the time. Curious little facts about complex transcendental functions and other such things that might come in handy in a calculus course are missing.

Math Education and What's Wrong with It

On its own terms, math education in the U.S., despite decades of sincere efforts by platoons of dedicated teachers, is pretty clearly a failure. At the simplest level, we do a worse job than any other industrialized country. For more complicated reasons, results on standardized math tests inside the U.S. (not compared to other countries) have drifted downward for nearly 30 years.

Some people blame this trend on "new math" and other educational experiments. It's probably true that learning the binary number system never improved anyone's working arithmetic skills, but a careful survey of many years of test results shows that circumstances in the glorious old days of straight arithmetic drills weren't fabulously better.

I think the problem is that neither old-time arithmetic nor set theory are particularly interesting to most young students. Some kids actually like to memorize times tables, but most kids don't. If only the little rascals knew that set theory was setting their tiny feet on a path to a true appreciation of Lebesgue integral theory in an upper-division college math class, perhaps their hearts would beat faster.

Perhaps not. In this book, I propose to remedy the motivational problems of the traditional arithmetic class by concentrating mainly on issues that probably have caused you trouble in real life. As an added cultural literacy bonus, I'm also going to tell you what your teachers were hoping to accomplish in high school math. I can pick through all the material presented to you in 13 years of K – 12 education (kindergarten to senior year in high school) and pull out 300 or so pages that are actually going to make a difference. You probably haven't done a Euclidean proof lately, but I bet you've made car payments.

A Word about Calculators

Calculators are great. People were not meant to do compound interest problems using a blackboard and chalk, much less in their heads.

Yes, of course, it's sort of a disgrace to find yourself multiplying 2×10 on a calculator. But there's no need to feel ashamed at using one to multiply 17×395. Personally, I did not grow up with calculators. When I was a tiny tot, I mastered both the Chinese and Japanese abacus just for fun (I was a strange kid), and in fourth grade I took a college-credit course in scientific slide-rule use. When I was a university professor, my colleagues used to lament that replacing the slide rule with the calculator was the ruination of the student mind, and I've heard most of the other educational arguments against calculator use in classrooms.

These arguments are not only misguided in general terms, but they also don't apply here. If you're reading this book, you're probably an adult. Your arithmetic quiz days are over. You're entitled to use any means necessary to get the answers you need. Getting the right answer, not purity of method, is the point. If using a calculator is cheating, then I'm giving you permission to go ahead and cheat.

Using This Book

The chapters in this book are free-standing and independent, even the closely related chapters on credit card payments and interest. Just find the chapter you need and start reading. If your checkbook is always out of whack, read Chapter 1. If you keep losing money on lottery tickets, read Chapter 17. If waiters routinely give you dirty looks, read Chapter 16. I've anticipated your every need, and I'm here to help.

How This Book Is Organized

It isn't organized, really, because the chapters can be read in any order. But traditionally, . . . *For Dummies* books are built on a strict Parts structure, and this one is no different.

Part I: Personal Money

This part introduces, in several variations, the ways in which interest, payments, and so forth are computed. It also treats checkbook reconciliation, which, let's face it, is more a matter of "just doing it" than an arithmetic challenge.

Part II: Business Math

Instead of personal finance, this section is really about economics — in the marketplace, in government, and in insurance (which is more like lottery practice than checkbook work).

Part III: High School Confidential

The problem with high school math courses is that they're really designed as preparation for college math courses, which are designed as preparation for math in graduate school, which is preparation for a math research career. If you're reading this book, I expect that you are not a research mathematician. If you are, go study the financial derivatives market instead. In this part, I talk about the issues in algebra, geometry, and trigonometry that actually appear in daily life. You can enjoy these topics when you're not stuck facing a life-threatening math exam every Friday.

Part IV: Off Duty

Look at it this way: If I can talk you into playing blackjack once a year instead of buying a lottery ticket every week at the 7-Eleven (hey, they named it that even before you could gamble there!), you could have paid $200 for this book and it would have been worth it. Also, I hope to change your perspective on professional sports, both your understanding of how the games work and why or why not to bet. With the money you save, you'll be able to cruise the finest dining establishments, so in the interests of hard-working food service personnel everywhere, I also discuss tipping.

Part V: The Part of Tens

In this part, I collect both useful and fun bits of information into little lists. And hey, this is a math book — it's just made for lists.

Icons Used in This Book

A grand . . . *For Dummies* tradition is the use of little icons to mark sections. (Actually, this is *my* grand tradition, since I wrote the first computer book to use icons way back in 1984.) Here's what the icons in this book mean:

This icon indicates a shortcut or an easier way to do a computation.

If a little inattention can lead to a wrong answer, or if you're likely to lose money easily, I'll flag it with a warning.

Can't be helped. Every now and then, I'll have to explain the basis of some operation (usually financial), and a little bit of algebra appears. You can skip almost everything marked with this icon, if you must.

 If it's important for you to remember a bit of information, I'll stick one of these icons next to it.

 Believe it or not, math has quite a bit of entertainment value, and lots of numbers have strange family histories. This icon points out some of that fun stuff.

 You'll find practical examples of mathematical concepts marked with these icons.

Part I
Personal Money

By Rich Tennant

"THE SHORT ANSWER TO YOUR REQUEST FOR A RAISE IS 'NO.' THE LONG ANSWER IS 'NO, AND GET OUT OF MY OFFICE.'"

In this part...

The latest reports on consumer debt in the United States show that our school systems are failing to teach one math subject that actually has lots of impact in daily life: compound interest. Unless you live a cash-only existence, as an adult you are enmeshed in the world of credit-card interest, mortgages, and considerations of what's tax deductible and what's not. Thus I have postponed the stroll down Mathematical Memory Lane to the third part of the book and will now plunge right into down-and-dirty topics.

The fact is, unless you were a business major, it's unlikely that you learned about any of this stuff in school. It's not like you took a test on Consumer Credit in college and flunked it — here in the land of opportunity, you get your first chance to flunk all your credit exams in real life. The study material is printed on the back of your credit card statements in tiny, pale gray letters. The kind of loan agreements you meet in life are famously difficult to read. And of course, there are the basics of the tax code. It's almost funny how much emphasis high school math places on the formula for the volume of a cone or a hyperbolic tangent while it nearly ignores the subjects of money and interest. Oh well, they don't teach much real French in French class, either.

Read the chapters in this part (or the other parts) in any order, depending on your own curiosity. You will find that the toughest math stuff called for here is the ability to multiply numbers by using a calculator. Start, if you will, with some topic that affects you personally. Read whatever you want, even if it's just the chapter on puzzles.

Chapter 1

Balancing Your Checkbook

● ●

In This Chapter

▶ Motivating yourself to maintain your balance correctly

▶ Easy steps for balancing checkbooks

▶ More checking details

▶ Approximation, or "good enough" checking

● ●

*O*ften, people don't actually finish reading books (this always comes as a dismaying surprise to authors). That's why I put this chapter on checking accounts first — if you read just this chapter and never get around to the chapter on trigonometry (Chapter 14), you'll still get your money's worth from *Everyday Math For Dummies.*

Here's some very good news:

You already know how to do your checkbook.

What I'm going to tell you will very likely make the process of reconciling your checkbook easier. But the fact is that you can already do the arithmetic, especially with the help of a calculator. You don't have to be scared of the process.

If you're reading this book, there's a good chance that you are one of the millions of people who glance over bank statements once a month but don't actually grind through the process of making sure that your checkbook total matches the bank's total for your account. I want to try to talk you into changing that behavior.

A Little Visualization Exercise

I understand completely. For years, I used to toss my bank statements in a shoe box and pretend that I was keeping track of my checking balance in my head. The issue wasn't arithmetic (I'm one of those people who bought calculators way back when they cost $300) — it was motivation. Until I started bouncing checks occasionally and the banks of California decided that this sort of negligence could be a gold mine (returned-check fees, or "bounce fees," increased four times in just a few years), I just didn't give the matter much thought. When I give you the speech about reconciling your checkbook, please don't imagine that some guy is smirking at you because his checkbook has been balanced to the penny since age 14. Nope, not me. This is a reformed sinner you're dealing with here.

I want you to picture a very specific scene: You go into your local bank branch to make a deposit. The teller looks at your name on the deposit slip and shouts, "Hey, the sucker's here!" The rest of the bank staff begins dancing around, waving $20 bills at you, and singing "What Kind of Fool Am I?" (I just put that part in because I don't like Anthony Newley.) You look around the bank and notice that a door near the safety deposit vault is open. Employees in overalls with pitchforks are piling up money beneath the "Overdraft Charges" sign. The teller, still laughing, hands you a little color catalog full of all the things you could have bought with the money that the bank charged you for bouncing checks. As you leave, the staff is pointing at you and jeering. You look down and notice that your shirt now neatly lists dates and amounts of all your overdraft charges.

Starting Simple

You want to avoid supporting your own bank in this grand manner. And it is relatively easy to avoid, too. You can keep your checking account in good order in just a fraction of the time you spend consulting the TV listings each month. That's why I propose to keep up the motivational approach in this chapter — you have the time; you own a calculator; and if you're bouncing checks, you're doing the equivalent of bouncing them on purpose. And it *never* pays to do so, no matter how broke you are. The bank makes sure of that.

Look at Figure 1-1, which shows a blank sample page of a checkbook register.

Check No	Date	Transactions	Transaction Amount	Deposit Amount	Balance

Figure 1-1: A sample blank checkbook page. Look familiar?

Some people use every line of the checkbook register to record actual checks. The preferred style is to use the gray zones for information, such as the purpose of the check (although going to a Safeway once a week and writing **groceries** in the gray space under every Safeway check is a bit ridiculous). Another tactic is to use the gray spaces for deposits rather than checks so that you can scan down the page and locate deposits quickly.

In real life, the gray spaces can be pretty useful as well as plentiful. When you get a standard box of checks and a check register, the register has enough space for a white and gray line per check, with room left over for your life story.

Figure 1-2 shows the simplest case imaginable — a brand new checking account with only a few checks written.

Check No	Date	Transactions	Transaction Amount	Deposit Amount	Balance
					500.00
101	9/1	Ramos Shoe Repair	20.00		
102	9/2	Safeway	110.00		
103	9/5	Time-Life Books	35.00		

Figure 1-2: Checking: a basic example.

Notice also that to make things simpler, I have used unrealistically rounded-off numbers for each check amount.

Now, inspired by the information that almost 15 percent of adults in America have never, not once, reconciled their checkbooks, I ask you to do the totals right here, writing the numbers in pencil right in this book. See? Nothing to it.

My guess is that people avoid total nonstop disaster by inquiring about balances at an automated teller machine (ATM) or by calling the bank almost daily. Please keep in mind, however, that the number you are given at the ATM or over the phone isn't your real balance since it doesn't include checks you have written that haven't made it back to your bank yet. Actually, some people get to be quite skilled at guessing balances, but doing so involves an unconscionable amount of background worrying. So please write in the totals for this example to get some practice.

Basic Checkbook Arithmetic

Now, doesn't that make you feel better? You went out and bought a book about math and did the first problem. Figure 1-3 shows the solution, and I bet you got everything right. Congratulations! What I'll show you next is a way to make all checkbooks work just like this example.

Check No	Date	Transactions	Transaction Amount	Deposit Amount	Balance
					500.00
101	9/1	Ramos Shoe Repair	20.00		480.00
102	9/2	Safeway	110.00		370.00
103	9/5	Time-Life Books	35.00		335.00

Figure 1-3: A checkbook in fine order.

Now, you had two things in your favor in this particular case. First, you had a starting point. It said right at the top of the page that the starting balance was $500, and the first listed check, 101, was the first ever drawn on this account. Second, the numbers themselves were easy. If you got paid $4,000 net income per month and all the checks you wrote were multiples of $100, checking wouldn't present much of a problem.

In Figure 1-4, you see where it all goes wrong in the world of checking. The top of the page contains a total, all right, but there's a daunting arithmetic problem. A huge fraction of the checkbooks in this great land of ours look just like this — checks pretty faithfully recorded, without a running total anywhere in sight. Another large fraction of checkbooks look like this example, only without a total at the top of the page.

Check No	Date	Transactions	Transaction Amount	Deposit Amount	Balance
					500.00
101	9/1	Ramos Shoe Repair	17.58		
102	9/2	Safeway	89.53		
103	9/5	Time-Life Books	24.97		
104	9/5	Electronics Today	17.95		
105	9/5	Pacific Electric	71.81		
106	9/5	Brite Cleaners	15.80		
107	9/6	Southside Saloon	39.00		
108	9/7	Home Hospice Foundation	40.00		
109	9/7	(cash at ATM)	80.00		
110	9/8	Molsberry's Market	66.48		

Figure 1-4:
A checkbook with some real numbers, including sales tax.

The problem is that once you skip the subtraction on just a few entries, you aren't going to catch up unless you sit down at the old kitchen table and make it happen sometime at night. That's right, I'm hallucinating.

Various fix-ups have been offered through the years, of which checkbook calculators probably made the most sense (but if you remembered to use the calculator, you could do your math right there in the store). Probably the fanciest electronic trick is a $300 checkbook calculator from Panasonic that actually prints the checks — it's like a portable accounting system! My guess is that in Japan, where consumer electronics are even a bigger deal than they are in the U.S. and where checking accounts are still something of a novelty (there you can carry large amounts of cash safely), this could be a hot item.

Southside Saloon??

The check examples in Figure 1-4 are thinly disguised entries from my own checkbook, except that about two years ago I decided to clean up my act and start maintaining current balance totals. There really is a Ramos Shoe Repair.

There really is a Southside Saloon, too, but it's probably not what you expect. I live in Wine Country in northern California, one of the most chi-chi, boutique-ridden, bed-and-breakfasted spots on the planet. Southside Saloon is owned by a famous chef from Sonoma, and it's a restaurant that specializes in oysters and different new-age cuisine delicacies. In this part of the world, if the sign says *Red Dog's Blood-and-Guts Workingman's Tavern,* the boys at the bar are usually wearing little tassel loafers and cashmere sweaters, and they "work" on cellular phones. The $39 shown in the check will get you about four glasses of Chardonnay and a respectable tip.

A Sporting Proposition

Look down the list of numbers in Figure 1-4, and look at the total at the top. Now stroll with me (I've *given* you this account!) over to an ATM cash machine located at the end of the block. The question I have for you is simple:

Do you think you can get $40 instant cash?

Well, do you feel lucky, punk? Want to bet me $100 that the transaction will go through? Oddly enough, I have been in this situation dozens of times, wandering off with someone to a cash machine only to find that my companion really didn't know his or her account balance with any certainty, often with embarrassing results.

Figure 1-5 gives the actual total for these numbers. But I'm writing this book, so of course I already calculated the answer. I wouldn't take your $100 under these unfair circumstances.

The $40 withdrawal is just going to miss. At least at an ATM, you can find this out for free. If you write a check for $40 at this point, the bank will have a party at your expense. You may have an account that makes automatic transfers, charging you for the privilege. Or the bank may just go ahead and bounce the check and let the merchant to whom you wrote it charge you for the returned check while using it for cash register decoration, to the amazement (or amusement) of your friends.

Check No	Date	Transactions	Transaction Amount	Deposit Amount	Balance
					500.00
101	9/1	Ramos Shoe Repair	17.58		
102	9/2	Safeway	89.53		
103	9/5	Time-Life Books	24.97		
104	9/5	Electronics Today	17.95		
105	9/5	Pacific Electric	71.81		
106	9/5	Brite Cleaners	15.80		
107	9/6	Southside Saloon	39.00		
108	9/7	Home Hospice Foundation	40.00		
109	9/7	(cash at ATM)	80.00		
110	9/8	Molsberry's Market	66.48		36.88

Figure 1-5:
A $40
near-miss.

Reality Time

You don't want this nasty bounced-check business to happen to you. *I* don't want it to happen to you, and our only connection is through this book. So it's time to examine your own checking account to see what needs fixing.

The main problem is that if you have let things slide, you don't have a starting point for reconciling your account. If you had a starting number, you could do all the subtractions necessary to get a balance — assuming that you had been recording your checks.

Unfortunately, you don't have a real balance, and the bank doesn't either. The bank shows a current balance, but it doesn't know how many checks you have floating around in the mail or elsewhere, so its balance is no better than your guess. The catch is, of course, that the bank has the last word.

Getting to the point

You can get to a starting point in several ways:

- ✔ If you're lucky, your bank prints an instant statement at an ATM. This isn't the best starting point, however, since it usually reports a fairly limited number of recent checks.

- ✔ Your bank probably provides a phone service that lets you call and see what the bank is reporting for your balance and also find out which checks have cleared. If you can use this service for a long enough history of checks, it works pretty well.

✔ At the end of the month, you get a bank statement with a listing of a month's worth of checks. This is probably your best bet for finding a starting point.

The monthly statement is the best starting point if you're completely lost because it includes a long list of checks. The lurking joker in the deck is always the check that you wrote two months ago that hasn't been posted yet at your bank. My own personal record is a check for $175.19 to a bookstore that took *three months* to clear, probably because the store owner went out and partied all night after finally unloading a three-volume *Dictionary of Byzantine Civilization* on someone and forgot the check in the pocket of his tuxedo. But if you have lost track of the real checking total, you can't get a true balance unless you're sure that all your old checks have cleared.

What the statement means

Your bank statement has two parts (a front and a back), not only because it's a two-dimensional piece of paper but also because it lists two kinds of information. On the front of the statement is a listing of checks cashed prior to the date listed on the statement. Figure 1-6 shows a little segment of such a listing. Notice that one of the items in the listing has an asterisk next to it, indicating that the number sequence of checks has been interrupted.

Figure 1-6:
Checks out
of numerical
order.

CHECK	DATE PAID	AMOUNT
766	11/15	36.00
767	11/15	31.46
770*	11/16	115.12
771	11/16	58.38
772	11/17	61.50

There are three possibilities at this point:

✔ A check got destroyed (you may have written it incorrectly and torn it up in the store).

✔ The check has already cleared and appeared on last month's statement. If you have that statement handy (that is, properly filed in the correct shoe box), you can check. If you look in your own check register and see that the check was written to a local place that typically cashes checks at your bank the very next day, the check probably has cleared already.

✔ The check is flying through never-never land but will indeed land some-day. You need to consult your check register for the amount.

On the back of your monthly bank statements, there are usually some well-intentioned forms for reconciling your account. They are about as inviting as your W-4 statement from work or the front page of your income tax forms.

Figure 1-7 shows what the bank is trying to tell you.

My own checkbook balance reconciliation form:

The bank thinks I have this much:

But they haven't seen these checks yet:

#	$
#	$
#	$
#	$
#	$
#	$

So the total is really this much, after check number _____ .

Figure 1-7: A reality-oriented reconciliation form.

The bank is saying to you, "Look. As of the last check we show on the front of this document, we think that you have this much. We have taken out our monthly fee. We have also subtracted all those little charges for using an ATM at a different bank. You may have forgotten about them, but we never do. The only thing missing for a correct balance is that we obviously don't know anything about missing checks or checks written after the statement date."

So here's what you do.

Reconciliation part one: up to the statement date

1. **Take the bank's total and write down the last check number. Put each number in the right space.**

2. **Look back at the bank's check listing and note which checks are missing (an asterisk or some other indicator will tell you when the number sequence has been interrupted).**

3. **Decide whether the checks missing in the bank's list have already been cashed, are really missing (destroyed), or just haven't landed yet (that is, by statement time).**

   The surefire way to resolve this is to keep all your old bank statements since, statistically, the first case is the most likely.

4. **If the checks missing from the list really are outstanding, write them in the little spaces.**

5. **Get out your calculator and do the subtraction.**

   Usually, there will be only a handful of checks that you need to track down to make this adjustment (see Figure 1-8).

My own checkbook balance reconciliation form:

The bank thinks I have this much:

$603.51

But they haven't seen these checks yet:

#455	$ 24.30
#517	$146.00
#518	$ 31.15
#519	$ 21.87
#520	$ 59.03
#	$

So the total is really this much, after check number _#520_.

$321.16

Figure 1-8: Typical statement balancing involves just a few checks.

Congratulations! Following these steps brings you into agreement with the bank for the statement date. Doing so also gives you a starting point for next month.

By the way, if you shop for bargains, you will find that you can get a *printing* calculator with the charges you save from avoiding just one bounced check. If you really, really hate this sort of activity, it helps to buy a calculator that you actually like (maybe you think it looks cool) as motivation. Of course, the uptown way to do any of this is to get a copy of the computer program Quicken or, alternatively, to do your calculations in a spreadsheet. But if you haven't bothered to do your totals for months and hate this stuff anyway, I bet I can't talk you into a computer solution for your checking problems.

Also, reading your statements from time to time is quite informative. You may find that all sorts of funny little charges are applied to your account. One of the strangest categories is nonbank ATM charges. For example, do you get charged a fixed amount for using your ATM card at the ATM unit at your supermarket checkout? If so, what's your motivation for not writing a check?

In the old days when there were just checks, you didn't have to keep track of all the little bank-bites ranging from 20 cents to two dollars or more — another reason that keeping an account correct to the penny is fairly hard to do. One more reason is direct deposit of payments, which sometimes toggles by small amounts with changes in tax law. By using the statement at the end of the month, you get a starting point with the actual correct-to-the-penny amount. (It's like Hansel and Gretel finding the trail of bread crumbs in the woods.)

Reconciliation part two: up to right now

You now have a check number and a correct balance. That is, after that last check cleared at the bank, you had the balance that you just calculated. Please note that the statement date is probably a week old. It's time to bring you up to date with your most recent check.

1. **Using the same form (go make a bunch of photocopies of Figure 1-7 before you use it!), make a note of your starting balance.**

2. **If you made any deposits, add those to the balance and enter *that* number in the box.**

3. **Now go look up all the checks that you have written since the last check on the bank statement. Add them all up to get a single number total. Put that number in the checks column and subtract.**

There you are. You've got a total! It's time for a little heart-to-heart talk.

One thing you can see from this exercise is that if you fail to write down checks, there's no way to find your balance. The bank will get all your checks sooner or later, but *you'll* never know what's going on yourself at any given moment. As far as that goes, I recommend printing the check transactions instead of doing them in standard handwriting. It's just a little habit that makes you write the numbers more neatly.

Also, I know that many times people don't check for a total simply from fear. The problem is that whether or not you want to know if you're in trouble, the numbers won't forgive you. You're better off knowing that you're going to be $50 in the hole *a few days ahead of time* instead of getting a little letter from the bank. Maybe you can borrow a bit from another account, spend an evening rolling quarters, or hold off on a payment that would put you into bounce mode. Anything beats forking over a pile of overdraft charges to the bank.

The Last, Best Hope for Checkbook Problem Children

The problem with the first sections of this chapter is that you could have taken all that advice anytime. There must be 20 well-intentioned magazine articles a year about checkbook balancing. The back of your bank statement tells you the same thing. You can buy computer programs that basically play back to you the same set of steps that everyone else tells you to carry out on paper.

I would modestly suggest that I have outlined the simplest version of this process ever to see daylight, but . . .

You're still *not doing it, right??!!*

I'm not trying to be insulting — I'm your friend here. But *you* know who you are. Check for the warning signs:

- ✓ You've got a checkbook register in which the balance column never has a balance.

- ✓ Every three months or so, you walk up to an ATM and ask it for money, and it tells you that you have –$15.91 in your checking account.

- ✓ You go away for a weekend, and when you come home, you find six identical little envelopes from your bank in the Saturday mail, each one telling the same tiny sad story about the bank covering a piddling overdraft and charging you $10 or $20 to do it.

Not a pretty picture. I feel your pain. It used to happen to me, too.

And the problem won't go away by itself. I pointed out to a friend of mine that his overdraft charges in 1994 alone would have supported a trip to Hawaii — and rather than ask me for helpful checkbook advice, he encouraged me nevermore to dwell upon this possibility (imagine!). What I'm going to suggest is that you do a correct checkbook reconciliation from time to time and then keep accurate day-to-day track with a simpler arithmetic procedure.

Checkbook psychology

Look. You can probably do arithmetic. You certainly own a calculator, since you get several chances in the course of a year to acquire one for free. So what's wrong?

I think I know what the problem is. See if you recognize yourself in the following scenario:

> You're in a long line at a supermarket. Everyone in front and in back of you is grumping up a storm about delays. They've all read the tabloid headlines that Cher is actually a robot operated by *two* small aliens and that Bigfoot has presented a plan for world peace to the U.N. The crowd is therefore impatient and unamused. Now it's your turn. The checker rings up a few bags' worth of stuff, and the total is $87.19.

Question 1: Are you, or are you not, going to stand there in line doing arithmetic as the rest of the people in line stamp their feet and moan? Of course you're not — I'm talking about *you* here.

Question 2: Are you going to do your subtraction chore as soon as you get home and put the groceries away? I believe "yeah, right!" would be approximately the correct vernacular response. You don't even need the involuntary condescension that would be implied in a Question 3.

Don't lose hope. There's a way out of this problem, but it's a radical departure from the standard advice for checkbook management. For the rest of this chapter, I'm going to assume that you will probably never again have a checkbook balance that's correct to the exact penny. But you won't care. You won't get into trouble, either, and if you bounce a check, you'll at least be doing it on purpose instead of by accident. This is a last-resort, only-if-you-have-to procedure, but it works.

Getting it almost right

I'm going to show you an everyday math trick for checkbook management, but first I need to demonstrate the math involved. So far, the checkbook examples have employed lists of 20 or so numbers on a page. Adding up these numbers by hand would be a modestly challenging task, and adding them up on a calculator is even a bit of a nuisance. There must be an easier way.

Pennies? Who needs pennies?

There *is* an easier way, and it involves accepting a slight loss of accuracy. How much accuracy? You get to decide. If you take one of the checkbook pages presented earlier (refer to Figure 1-5) and simply round off every entry to an even number of dollars, you get a balance of $36 instead of $36.88. The rounding procedure is simple: If the cents part of the figure is $.50 or less, you count it as zero, and if it's $.51 or more, you round up to the next dollar. After all this rounding, notice that the total is within a dollar of the correct figure.

The point, for this purpose, is to notice that if you keep track of your checkbook just in terms of whole dollars, your recorded balance will be very close to the real balance as long as you bring it back in line every few months.

This trick works because the differences between the rounded numbers and the true numbers cancel — about half the differences are positive and half are negative. If you want to be extra careful about never bouncing a check, deduct an extra dollar every ten checks or so.

If you don't keep track of pennies in your checkbook, you cut your arithmetic workload almost in half. Hang on — I have an even more radical proposition than just dropping pennies. I'm going to keep cutting the arithmetic workload until you're willing to do the numbers right there in the supermarket line.

Rounding for the cautious

If abandoning the idea of an exact total really makes you nervous, just round the numbers *up* every time. Even $11.04 gets rounded up to $12. It's not as accurate as the real rounding procedure, but it's got insurance built into it. You do less arithmetic, and your checking account always has just a bit more money in it than your account shows. The problem is that you don't know how much more, and you might start feeling optimistic. Remember, you're reading this chapter because optimism may have caused problems in the past ("Yeah, I'm pretty sure that check should be good . . .").

Always rounding up to whole dollars is the equivalent of establishing a miniature secret savings account into which, on the average, you deposit $.50 every time you write a check. After 40 checks, you have a built-in cushion of $20!

Radical rounding

If you can handle it, you can take an even more radical approach to rounding. Try rounding to the nearest $5 — some people like doing so because years of making change with dimes and nickels and quarters seems to make people better at arithmetic problems where all the numbers end in five or zero. Figure 1-9 shows the results.

You write the real figures in the checkbook (writing them in the gray area is probably easier) and write the rounded figures in the Transaction Amount column. Note that, following this trick, you get $35 instead of $36.88. Given that you probably didn't know what all the other little charges from the bank were this month, this level of accuracy is not bad. To be extra careful, charge yourself an additional $5 after every 20 checks. (Technically, this trick prevents random errors from getting out of hand.)

Check No	Date	Transactions	Transaction Amount	Deposit Amount	Balance
					500.00
101	9/1	Ramos Shoe Repair --17.58	20.00		480.00
102	9/2	Safeway -- 89.53	90.00		390.00
103	9/5	Time-Life Books -- 24.97	25.00		365.00
104	9/5	Electronics Today -- 17.95	20.00		345.00
105	9/5	Pacific Electric -- 71.81	70.00		275.00
106	9/5	Brite Cleaners -- 15.80	15.00		260.00
107	9/6	Southside Saloon -- 39.00	40.00		220.00
108	9/7	Home Hospice Foundation --40	40.00		180.00
109	9/7	(cash at ATM) --80	80.00		100.00
110	9/8	Molsberry's Market -- 66.48	65.00		**35.00**

Figure 1-9: Rounding checks to the nearest $5.

So what happens if you round to the nearest *$10*? Amazingly (at least until you think about it for a while), this rounding procedure gives perfectly acceptable accuracy on a month-to-month basis. Look at the results in Figure 1-10.

Check No	Date	Transactions	Transaction Amount	Deposit Amount	Balance
					500.00
101	9/1	Ramos Shoe Repair -- 17.58	20.00		480.00
102	9/2	Safeway -- 89.53	90.00		390.00
103	9/5	Time-Life Books -- 24.97	20.00		370.00
104	9/5	Electronics Today -- 17.95	20.00		350.00
105	9/5	Pacific Electric -- 71.81	70.00		280.00
106	9/5	Brite Cleaners -- 15.80	20.00		260.00
107	9/6	Southside Saloon -- 39.00	40.00		220.00
108	9/7	Home Hospice Foundation -- 40	40.00		180.00
109	9/7	(cash at ATM) -- 80	80.00		100.00
110	9/8	Molsberry's Market -- 66.48	70.00		**30.00**

Figure 1-10: Rounding to tens.

So how does this sound to you? You get your checkbook in order once every two or three months to give yourself a starting point. After that, you only have to work with subtractions involving ten dollars. Those earrings for $49.23 after tax . . . 50 bucks. A check for $4.22 . . . zero. Now when you're standing in line at the supermarket and the total is $89.17, you will be subtracting $90 from, say, $400. Hey, you can do that one while the whole world waits and watches. Tell 'em to look at the tabloid article about Madonna bearing Barney the Dinosaur's love child to pass the time while you casually subtract a single digit right on the spot.

"But," you may object, "this is *crazy*! It's just too weird rounding off numbers this way." Well, no, it's not; the procedure works, and you don't need a calculator. If you do a proper reconciliation every 50 checks or so, you'll very seldom be more than $25 lower than your own records show, half the time you'll have a bit *more* money than you think, and typically you'll be within $5 of the exact balance. You can also just charge yourself $10 every 20 checks for using this procedure, just to be safe.

Isn't that better than being totally lost most of the time? Isn't it better than being $132 out of whack because you never do the subtractions? Doesn't it beat getting all those smirking little notices from the bank as it pockets outrageous fees? If you keep a checkbook register like Figure 1-10, you'll have all the correct documentation, you'll have the advantage of arithmetic you can do right when you write your checks, and you'll always show a balance that's "good enough" so that you don't get into trouble.

Actually, I have another motivation here. If I can recast the arithmetic so that you're willing to do it, the whole subject of checkbook balance reconciliation becomes a sort of game instead of a chore. If you know that you're within a few dollars of the correct balance, you'll *want* to see what the real to-the-penny total is when your bank statement comes. This will be a repeated theme in this book: If I can get you involved, you'll find that you can do all sorts of amazing things.

What about Quicken?

If you at least have the patience to type in all the entries in your checkbook into a computer-based checkbook manager, you can get the arithmetic done for you automatically. The program Quicken from Intuit Corp. is available for PCs and Macintoshes and is the clear market leader.

A Quicken screen (see Figure 1-11) looks pretty much like a check register page, but it can do all sorts of other tricks. It can pay bills automatically with electronic bill-paying services, link your checkbook to credit cards, and pass all your financial information along to income tax software. It's also quite easy to learn.

The only catch is that you have to be willing to enter the information — if you don't keep Quicken up to date, it can't help you. But if you are willing to make the minimal effort of typing in the data (probably 30 checks a month in most cases), Quicken will take care of all your checkbook balancing and other transactions.

For more information about using Quicken, read *Quicken For Windows For Dummies*, *Quicken For DOS For Dummies*, or *Quicken For Macs For Dummies*, all from IDG Books Worldwide.

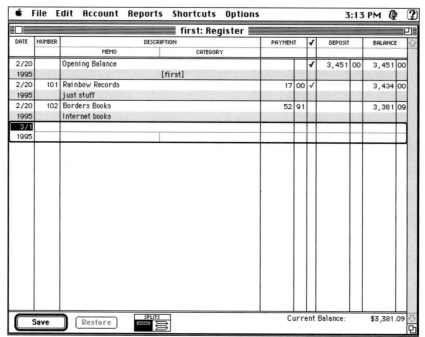

Figure 1-11:
A sample
Quicken
screen.

Chapter 2
Time Payments and Interest

• •

In This Chapter

▶ Understanding compound interest

▶ Payment tricks

▶ The payment formula

▶ Getting on the right side of the process

• •

*R*each into your pocket and pull out a $20 bill (if you don't have one, any amount of money will do). Put it on a table in front of you and look at it. It may seem to be resting there peacefully, but it's actually undergoing considerable turmoil. You can read a longer description of these tumultuous processes in Eric Tyson's *Personal Finance For Dummies* (IDG Books Worldwide, Inc.), the best-selling companion volume in this series.

First, the money is shrinking as you watch it, because most modern economies have a small but not ignorable inflation rate. This steady inflation is a quite recent phenomenon. In many periods of American history, *disinflation* (or *deflation*) was normal — prices of crops and land, for example, would sink like a stone as farmers and speculators alike were ruined. Under those circumstances (which you are unlikely to see in a carefully managed modern consumer economy), your money on the table would be growing.

Second, the $20 bill is depreciating compared to the same money in a time deposit. Twenty dollars as a piece of paper lying on a table isn't earning any interest. However, the same $20 in a bank savings account or a certificate of deposit is earning some interest. In fact, $20 in the bank will earn a fraction of a cent in interest overnight, while the $20 on the table will just sit there.

When you buy a car or a refrigerator on a payment plan, the lender from whom you borrow has to take all this time value of money into account. In addition to the sheer joy of selling you a car, the lender gets to decide what the money is worth over time. If the lender is sending you home with a refrigerator that would be worth $500 cash right now, he or she has to take into account how much extra to charge you because the $500 won't be collected today.

Surprisingly, the math for compound interest turns out to be fairly fancy stuff. It always amazes me, on the frequent occasions when I think about the standard math curriculum in high school and in college, that you can be required to spend months laboring over set theory and nearly a year constructing geometric proofs (presumably so that Greeks in ancient Alexandria wouldn't be ashamed of you) and spend almost no time on the one math topic that really impacts daily life, compound interest. Somewhere out there, I'm sure someone probably remembers that

$$\cos^2 A + \sin^2 A = 1$$

but doesn't understand why consolidating debts for a lower monthly payment at a higher interest rate is not such a hot idea. Because the aim of this book is to plug up any holes in your education, I'll tackle trigonometry, too. Right now, though, I'm going to discuss compound interest and then elaborate on this theme over the next few chapters.

Annuities

With annuities, you are on the *winning* side — an *annuity* is a situation in which someone pays *you.*

If you read a lot of late Victorian novels, you probably read about people getting annuities. The situation usually goes something like this: Some young person receives an inheritance, say 10,000 pounds, and because people were smarter back then, this rather large Victorian fortune is deposited in a bank as an annuity account. Under the terms of the annuity, the money earns interest, but a payment comes at a fixed date every year. With an interest rate of a few percent but with a correspondingly modest annual payout, the annuity provides lifetime income.

This is the opposite of an installment payment plan, where you provide the income payments to a bank or lender because the bank or lender left some money or goods with you. Although this situation is the opposite in terms of money flow (in versus out), the math is the same.

A literary aside

For late Victorian adventures with an annoying level of attention to money details, I recommend the works of George Gissing and George Meredith. Near the turn of the century, Meredith was almost universally acknowledged as a true immortal and one of a handful of really great English novelists. I would love to hear from readers who have read a single work by this author outside a classroom course. There is, of course, the terrifying suggestion in all this that some novelists with great reputations today will be utterly forgotten in 50 years or so. Of course, for sheer intensity of money references in literature, no one can beat Balzac. Read *Père Goriot* and you'll start burying quarters in jars in the backyard.

Lottery! A Sort of Annuity

As a familiar example of an annuity, you really get an annuity (doled out in monthly payments) rather than a huge wad of dough when you win a state lottery. For example, suppose you win $2,000,000. The state divides this sum into 20 years of payments, giving you $100,000 per year.

Because the state has the whole two million on deposit somewhere, it can get at least five percent interest and thus can make your lottery payments without actually using up any of the original prize money. In other words, the state can usually make your lottery payments on interest alone. By the time federal and state withholding taxes are taken out, you should be seeing payments of about $5,200 per month. Not bad, but if you start making payments on a house in Beverly Hills and buy a Rolls Royce, you're going to need a day job, lottery or not. By the way, read Chapter 18 on gambling before you start daydreaming about the Rolls.

Figuring Interest Yourself

Here's an example to think about as a starting point in discussing interest. I'm taking the upside-down version of a credit plan first because the math is a little easier to follow. Here, you deposit $100 on the last day of every month. You deposit it someplace where you get interest, too. The question is this: How much is your whole set of deposits going to be worth on the last day of this one-year plan? So that you have a concrete framework to think about, suppose the first month of this plan is January and the last month is December.

Now, if there weren't any interest, the problem would be simple. You make 12 deposits of $100, so at the end of the year (on December 31) you have

Total amount in plan = $12 \times \$100 = \$1,200$

But in this case, all your deposits are earning interest. And they're earning different amounts because each deposit has been sitting in the bank for a different number of months.

Analyze them one month at a time.

The December deposit is just the $100 you put in at the end of the month. When you deposit it on December 31, it's still worth just $100.

The November deposit has accumulated one month's interest. Just to make things spectacular, suppose you found a bank that will pay you 18 percent per year. Right now (in 1995), you can find plenty of banks that will charge you that much interest on a credit card, but they're not paying out at that rate these days. If interest is paid monthly, then you get

Monthly interest = Annual interest/12

or

1.5% = 18%/12

The monthly interest in this arrangement is 1.5 percent.

How much is the November deposit worth on December 31? Because it has been sitting in the bank for a month, it has picked up one month's worth of interest. That means it's worth $100 plus another 1.5 percent of $100, which is

1.5% of $100 = $(^{1.5}/_{100}) \times \$100 = \$1.50$

This formula just uses the definition of percent. A percent is a fraction of the number 100. Just to take two examples, 8 percent is the number

$^{8}/_{100} = 0.08$

and 12.5 percent is the number

$^{12.5}/_{100} = .125$

Now, in total, the November deposit is worth

November = \$100 + \$1.50 = \$101.50 = \$100 × 1.015

So the November deposit is worth \$101.50, which is just \$100 plus the 1.5 percent interest.

That wasn't so tough, but what about the deposit from October? It has accumulated interest for two periods. Because you're dealing with compound interest here, you have the "interest factor" of 1.015 being applied twice.

The October payment, after sitting around in the bank for a month, is worth

November value = \$100 × 1.015 = \$101.50

when you apply the results of the preceding formula. Since this is the value at the end of November, the value at the end of December will be the same as if you had deposited \$101.50 at the end of November and then applied the interest to that amount. You are left with a December value (for the original October deposit) of

December value = \$101.50 × 1.015 = \$103.02

So on December 31, the value of your \$100 from the end of October is \$103.02, rounded to the nearest penny.

You follow the same procedure for each month, applying interest on top of interest, just using a monthly interest factor. The biggest value is the one for the January deposit, to which the interest factor is applied 11 times, accumulating a little bit more every month.

What does all this amount to after a year? You can see the results in Figure 2-1, a table that summarizes the preceding computations.

The table is arranged from the most recent payment at the top to the oldest payment at the bottom. There's also a total showing that, instead of having \$1,200 in the bank on December 31, you have \$1,304.21. The interest brings you a bit more than \$100.

Depositing Equal Payments for a year				
		annual interest in %=		18
		monthly interest		0.015
last pmt.	$100.00	1.0000	$100.00	
	$100.00	1.0150	$101.50	
	$100.00	1.0302	$103.02	
	$100.00	1.0457	$104.57	
	$100.00	1.0614	$106.14	
	$100.00	1.0773	$107.73	
	$100.00	1.0934	$109.34	
	$100.00	1.1098	$110.98	
	$100.00	1.1265	$112.65	
	$100.00	1.1434	$114.34	
	$100.00	1.1605	$116.05	
first pmt.	$100.00	1.1779	$117.79	total> $1,304.12

Figure 2-1: Compound interest on regular payments.

Paying Interest on Consumer Loans

If you deposit money at some reasonable rate, the money is out there working for you, even while you sleep. However, if you owe money, your debt is working against you while you sleep. When real interest rates (see the following sidebar) are high enough, depositors have an immense real-life advantage over borrowers.

Here is an example in detail. Suppose you borrow $1,000 at 20 percent annual interest and start paying it back with $100 payments every month. What will be your payment total when you're finally done?

Note that there is nothing unusual about this arrangement. Plenty of organizations are willing to charge you 20 percent annual interest, $1,000 will get you a not-very-spectacular wide-screen TV, and the only bit of surprise is that you are willing to make payments of $100 (the minimum payments would usually be lower).

The story is outlined in Figure 2-2. At the end of the first month, you pay your $100. The problem is that you don't owe $1,000 at the end of the first month — you owe $1,000 plus one month's interest. The monthly interest rate is

Monthly rate = Annual rate/12

so it's

Monthly rate = 20%/12 = 1.667%

For the first month, the amount of interest is $16.67. After you make your payment of $100, interest starts being charged to the amount $916.67, getting ready for next month's payment.

You have, of course, already concluded that you will be paying more than $1,000 on your $1,000 loan. How much more? Well, it looks from the table in Figure 2-2 that you're making 11 payments of $100 each, which is $1,100, with a last little blip of $3.04 at the end.

Making Payments Against a $1000 Debt			
Annual Interest =	20	**Monthly Interest =**	0.0166667
Payment=	$100.00		
		Factor	Balance
	$1,000.00	1.0166667	$1,016.67
	$916.67	1.0166667	$931.94
	$831.94	1.0166667	$845.81
	$745.81	1.0166667	$758.24
	$658.24	1.0166667	$669.21
	$569.21	1.0166667	$578.70
	$478.70	1.0166667	$486.68
	$386.68	1.0166667	$393.12
	$293.12	1.0166667	$298.01
	$198.01	1.0166667	$201.31
	$101.31	1.0166667	$102.99
	$2.99	1.0166667	$3.04

Figure 2-2: Interest charges on regular payments.

This extremely simple example nonetheless illustrates most of what you need to know about making payments. You make the payments, and the lender charges interest on the unpaid balance. Two little modifications make real-life consumer loans a little less attractive than this example.

First, real loans have a provision for partitioning your payments into an *interest* part and a *principal* part. The loans are set up to provide the lender with more interest in the earlier payments and less interest in the later payments.

Second, when inflation rates are low, the lender is usually in no hurry to collect, offering you lower monthly payments. The lower payments last longer, the lender collects more interest, and making the monthly payments is "easier" for you. The payments are lower, but you make lots more of them for a much larger total in the end.

"Real interest" and the great game of Life

Interest isn't the only factor in the time value of money. For example, if the interest you earn on a savings account is 5 percent but the inflation rate is also 5 percent, the real value of the money in your account is standing still in purchasing power.

As this book is being written, consumers can find themselves on the paying end of one of the worst deals ever offered legally. Back in the late 1970s and early 1980s, the inflation rate in the U.S. went into double digits for the first time since World War II. So that banks and savings associations did not utterly collapse, they were allowed to raise interest rates rapidly and write lots of new loans at the new higher rates. Of course, consumer interest rates (credit cards and others) were raised even higher than the rates on home loans.

Now inflation rates have backed down to a few percent per year, but credit card interest rates are still, in many cases, back up at dizzying levels (20 percent or so). This represents the highest "real interest rate" ever seen outside mob loan-sharking. If you have, for example, a big department store charge card balance, you need to start thinking about getting out from under this typically raw deal. As a little math exercise, look up the so-called prime rate, the latest reported interest rate, and the rates on your payment statements from any time payments.

"E-Z" Payments?

Let me make this last argument a bit more explicit. Figure 2-3 shows the same loan, at the same interest, only this time you make fabulously easy payments, a mere $16.50 per month, instead of $100. A mere $16.50 per month for a big-screen TV — why, that's less than basic cable TV service in most locations.

Making Payments Against a $1000 Debt			
Annual Interest = 20		**Monthly Interest =** 0.0166667	
Payment = $16.50			
		Factor	Balance
	$1,000.00	1.0166667	$1,016.67
	$1,000.17	1.0166667	$1,016.84
	$1,000.34	1.0166667	$1,017.01
	$1,000.51	1.0166667	$1,017.18
	$1,000.68	1.0166667	$1,017.36
	$1,000.86	1.0166667	$1,017.54
	$1,001.04	1.0166667	$1,017.73
	$1,001.23	1.0166667	$1,017.91
	$1,001.41	1.0166667	$1,018.10
	$1,001.60	1.0166667	$1,018.30
	$1,001.80	1.0166667	$1,018.49
	$1,001.99	1.0166667	$1,018.69

Figure 2-3: Comparing "easy" payments with larger payments.

The catch in this payment arrangement is, as you can convince yourself with a brief study of the table, that the payments never end. That's *never,* as in eternity, the age of the universe — you pass the debt on to your kids. The first month's worth of interest is $16.67; you pay only $16.50, so you start the next month owing a tiny bit more than $1,000. And thus you never catch up.

Real-world payment plans tend to pick a point somewhere between the relatively aggressive schedule ($100 = pay it off in a year) and the neverland schedule ($16.50 = final payment in heaven). If the lender wants the money back in three to five years, obviously you will pay something between $100 and $16.50 per month on this $1,000 loan at 20 percent. How is that payment amount computed in real life?

The Payment Formula

I produced the little tables in the first three figures with a computer spreadsheet, which lets you do all sorts of calculations in a table format. Every spreadsheet since VisiCalc of the mid-1980s has also featured a built-in function for computing payments. (For a painless introduction to the most popular spreadsheet, I recommend the *Excel 5.0 Starter Kit,* by — ahem — me.)

Because the formula for computing payments is the linchpin of modern civilization, the tie that binds us all together in a vast commonwealth of credit ratings as a measure of human value, it was, of course, worked out long ago.

The following lines show the main mathematical source of the payment formula — they come from the math program Mathematica, which can not only do algebra but also can do everything you ever encountered in school math (more on this subject in Part III).

$$S = 1 + a + a^2 + a^3 + a^4 + a^5 + a^6 + a^7 + a^8$$

$$Y = (1 - a); S = (1 - a)(1 + a + a^2 + a^3 + a^4 + a^5 + a^6 + a^7 + a^8)$$

Expand Y:

$$Y = 1 - a^9$$

In the course of computing the results in one of the columns in the spreadsheet, I multiply some amount by the interest rate two, three, four, five times, and so on for a year. The interest on the unpaid balance is *compounding,* meaning that interest is charged on interest. If you have a long string of numbers with repeated multiplications of the same factor, you can add them all up with an algebraic trick (I give the hard-core details in Chapter 12 — this trick works for the values of the variable *a* that occur in interest problems) and get this result:

$$(1 - a^n)/(1 - a) = 1 + a^2 + a^3 + a^4 + a^5 + a^6 + \ldots + a^{(n-1)}$$

You don't really need to know this, except that it explains where the calculator companies get the following odd-looking formula. If you look through books about finance, you see this formula, which is really no more than a one-line summary of the tables in Figures 2-1, 2-2, and 2-3. The fact is, the formula looks complicated because it has to account for the effects of lots of little repeated multiplications, making it difficult to simplify further. Fortunately, having seen this formula, you can now more or less forget about it and pay attention to results rather than to the machinery of calculation.

$$\text{Payment} = (i \times \text{Amt})/\{[(1 + i)^n - 1](1 + i)\}$$

Using the Formula to Calculate Payments

If your calculator has a key that says y^x on it, you can use the formula by plugging in numbers. In the formula, i is the interest per payment period. If the payment period is a month and the interest rate r is 9.5 percent, for example, you first are supposed to realize that 9.5 percent is the number 0.095 and then divide this number by 12 to get the *monthly* interest. *Amt* in this context stands for the amount of the loan, and n is the total number of payments.

But if you have the slightest interest in thinking about payments, you can get a financial calculator that has every variant of the payment formula. You can solve for payments, given the amount of the loan, the interest rate, and the number of payments. Or you can solve for the number of payments from the loan amount and the payment amount and interest rate. Four numbers are really in play here, and if you have three you can solve for the other one.

The least expensive financial (or business) calculators cost about $12 to $16. Meaning no disrespect whatsoever to your judgment as a consumer, I'll bet that you've wasted more money than that on things that did you a lot less good.

If you have a computer, an alternative is to use the PMT function in a spread-sheet. Even this is overkill in most circumstances. If you're connected to any sort of on-line service (such as America Online, Prodigy, or CompuServe), you will be delighted to find that calculator-type programs are available as freeware (at no charge).

Looking at a straight purchase plan

Here's an orthodox payment scheme right out of a Sunday newspaper. I'm going to check it with a little calculator program, selected because it looks like a pocket calculator on-screen (it's zCalc from Pixel City, 415-367-0808). The program, just like the calculator, simply uses the payment formula from the preceding section.

A particularly desperate Toyota dealer is offering a hot deal on trucks. He's desperate for three reasons: 1) The yen has increased so much in value compared to the dollar that Toyota prices are floating hopelessly upwards. 2) After being beaten up in quality-control ratings for decades, General Motors and Ford are doing many more things right. 3) The wild country boys in the tassel loafers and cashmere out here in Sonoma County at the Grey Hog Spit-on-the-Floor Saloon won't be caught dead in a Japanese truck anymore (hey, a man's gotta do what a man's gotta do).

The ad reads:

'94 4 × 4 Toyota Pickup

$339 per month plus tax

3.9% APR finan. 48 months

Selling price of $14,077.50

Total of pmts. $16,272

Now, what's really going on here? The ad gives an annual interest rate, a selling price, and a number of payments (48). Just in case you can't do it yourself, the ad also multiplies $339 by 48 to give a payment total of $16,272. Because of the quite modest financing rate (3.9 percent) you would be paying only about $2,200 in interest over the four years. If you plug the numbers in the ad into a calculator (see Figure 2-4), you get almost the same payments, but not quite. (Note that the calculator shows your payments as a negative number, meaning that you pay rather than receive the money.)

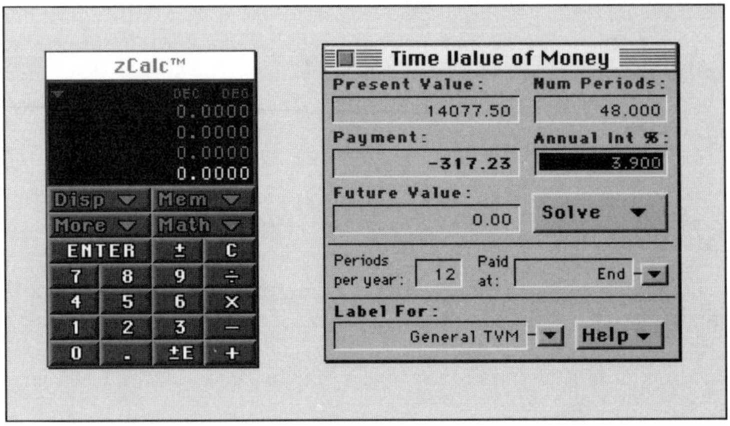

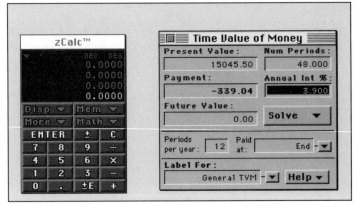

Figure 2-4:
Truck payments, almost.

The difference between the payments quoted in the ad and the payments computed by the calculator is interesting. Originally, I guessed that the $339 in the paper might be an amount that included sales tax on the monthly payment (7 percent tax brings $317 to $339). Upon calling the dealership, I found that the $14,077.50 does not include, apparently, a few more tweaks. The true amount being financed is almost $1,000 more, and the real computation appears in Figure 2-5.

Figure 2-5:
Truck payments, for real.

It's curious that in this age of mandatory full disclosure of financing information and tons of fine print in car ads, you still don't quite have all the facts at your disposal just from reading the print ad in a newspaper. In fact, the next section shows another example from the same dealer.

Looking at a lease

The same newspaper includes this ad for a lease on a Toyota Corolla:

$239 per month

$256.93, including tax, for 36 months

Total payments of $9,249.48

Residual price $7,135.96

This ad says that you can drive this Toyota around for $239 per month, and after paying out about $9,000 on a $12,000 car, you have to pay $7,000 more if you want to own the car outright. Yikes!

The way most lease plans work is quite simple, although the information given here isn't quite enough to let you figure it out. You have some information (the number of payments you make and the payment amount), but you don't know the actual sale price of the car, you don't know the interest rate in this payment scheme, and you aren't given much of a clue about the final ownership payment.

By reading more detailed ads (Saab and BMW really spill the beans) for auto leases, you can find out several things. Most leases, for example, just amount to a standard payment plan with payments lasting five or six years. The main difference is that you make only three years' worth of payments — that's why there's a big balance at the end. Another difference is that lease plans tend to be based on a higher interest rate than straight-purchase plans. You get lower monthly payments but are shelling out more money in interest and consequently own a smaller fraction of the car at the end.

The actual circumstances at the end of the lease (where you either finance the rest or turn the car over to the dealer and start another lease on a new car) suggest tinkering with the interest rate as a way to guess the lease structure. Figure 2-6 shows the results of some guesswork — it looks like the real car price is about $12,000, the interest rate is 12 percent, and the lease is modeled on a 72-month purchase.

The lease payments are computed as if this were a straight purchase with a six-year term. Because the interest rate is relatively high and the payments per month relatively small, you pay out $9,000 on a $12,000 car over three years and then have $7,000 more to go! Leasing has tax advantages if you use your car for business, and it makes sense if for some reason you need to have a relatively new car all the time, but it is not the lowest-cost method of vehicle acquisition.

TIP

There are, however, lots of variations on this sort of lease arrangement. In some cases, the interest rate is fairly modest, and the seller isn't trying to recoup much of the retail price because the depreciation over the lease term is small (that's why good lease deals are possible on the best luxury cars). The main points to note are simple: Find out how much of the price of the car will be paid off in the course of the lease. If it's a fairly small fraction, then most of what you're paying is interest.

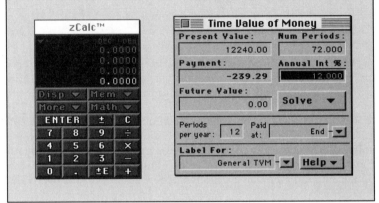

Figure 2-6:
Lease payments as a variation on standard payments.

A Car Payment Table

So that you can figure payments with a simple four-function calculator, here's a little table of monthly payments based on a $1,000 loan amount (see Table 2-1). If you are financing $12,000, multiply the payment by 12. Note that interest rates are often quoted in numbers like 7.9 percent. In using the table, just round this number to 8 percent. As one of the ongoing themes of this book, rounding off by small amounts gives you "good enough" results, which is a great advance on absolute cluelessness!

Table 2-1	Monthly Payments on a $1,000 Loan Amount				
	4%	*6%*	*8%*	*10%*	*12%*
2 years	43.42	44.32	45.23	46.14	47.04
3 years	29.52	30.42	31.34	32.27	33.21
4 years	22.58	23.49	24.41	25.36	26.33
5 years	18.42	19.33	20.28	21.25	22.24
6 years	15.65	16.57	17.53	18.53	19.55

Notice a few interesting points about this table:

- To restate the point, if you have a $13,000 loan amount, you just pick the term in years and the interest rate and multiply by 13 to get your payment.

- Each two points of interest (for example, going from 6 percent to 8 percent) takes the payment up by about a dollar. If you're figuring a 9 percent loan, then, you can just take a payment number halfway between the 8 percent payment and the 10 percent payment, which is in fact about 50 cents higher than the 8 percent payment. If I were trying to be impressive, I would call this *interpolation,* but it's such a common-sense thing to do that it hardly needs a Latin name.

- Small increases in payment amount make big effects on the final total of your payments. Looking at the entry for a six-year loan, the difference between 8 percent and 10 percent is, as I said, only a dollar. But it's a dollar times 72 payments, times the number of thousands of dollars in the loan. On a $12,000 loan, that's $72 \times 12 = \$864$. These are all short-term loans — this effect really gets magnified when you start computing mortgages.

Chapter 3

Credit Cards

● ●

In This Chapter

▶ Basic credit card math

▶ Payment plans and infinity

▶ Saving by borrowing?

▶ A plan for credit management

● ●

redit cards are amazing. On the one hand, they offer lots of convenience and even bring some security, as we live in a large country where robbery for cash is far from unknown. On the other hand, they often represent the equivalent of a free financial piece of rope with a noose-tying video included in the package. The standard middle-class household can get itself into serious money trouble with credit cards in no time. Credit cards are the financial equivalent of pain-killers; sometimes they're just the prescription for you, but without careful attention, you can slide into an addiction that really complicates your life.

This chapter, of course, reviews the basic math involved in using credit cards. My real aim, however, is to get you to look at the faint gray type on the back of your credit card statements, review the tables in this chapter, and then throw yourself on the ground in a cold-sweat agony of abject fear. Ladies and gentlemen, I ain't kidding. It's scary. And please don't think that I'm going to propose any easy way out of credit card trouble, such as paying off cards with a house refinance. That calculation is in the next chapter, but I can tell you before you get there that the refi "solution" isn't as promising as it might seem.

By the way, please consult Eric Tyson's *Personal Finance For Dummies* for a coherent strategic credit management plan. My job here is to "run the numbers" with you. At the end of some calculations, however, you'll be able to formulate some advice of your own. The lessons in this chapter are about as complex as "Don't stick that silver fork in the electric outlet . . . a second and third time."

Understanding the Fine Print

I'm going to review a sample credit card statement, which is actually a sort of combined rehash of VISA, MasterCard, and Discover statements. Conspicuously missing is the standard American Express statement — with the American Express card (*not* AmEx Optima), you pay off your balance at the end of every month. The cards I'm discussing here are *real* credit cards, the cards designed to let you buy things that you may not actually be able to afford.

Making the monthlies

If you can handle the eye strain, the back of your credit card statement has a series of fascinating statements to interpret.

Statement 1: Figuring the minimum monthly payment

The minimum monthly payment is the greater of 2 percent of the outstanding balance and $10.

This means, of course, that if you owe $635.10 on the card, your minimum payment is

$$payment = \$635.10 \times 0.02 = \$12.70$$

If you owe less than $500, your minimum payment is $10.

Statement 2: Calculating finance charges

To compute the Periodic Finance Charges yourself, refer to the finance charge summary and use the following equation for each Average Daily Balance shown there:

Average Daily Balance

$\times$ Number of Days in Billing Period

$\times$ Daily Periodic Rate

= Periodic Finance Charge

You may not be aware that the interest on credit card balances is typically compounded per day. Some credit card companies' accounting systems are fussier than others, with the computer noting the day of purchase and doing all this microcompounding on a day-by-day basis.

Also, you may be paying three different interest rates on different chunks of your overall balance. If your card allows cash advances, those usually accrue interest at the highest rate. There's also usually a difference between the interest charged on old balances versus that charged on the balance on purchases you made this month.

To simplify things, I'll just take the rate on unpaid, old balances. On the statements I'm using, the Daily Periodic Rate is 0.05425 percent. That makes an average monthly interest of

$$\text{monthly interest} = 30 \times 0.05425\% = 1.6275\%$$

That corresponds to an effective annual rate of nearly 21.4 percent if you start the year with an unpaid balance (although the card-issuing companies use the same numbers to arrive at a 19.8 percent rate by using a slightly different method for timing the compounding).

The paydown

As a first example, consider putting a balance of $2,000 on this credit card and making payments of $60 a month. This is a bit above the $40-per-month minimum payment the credit-issuing firm would want — I'll examine that particular case next.

The table in Figure 3-1 shows the results of the first set of payments. Basically, it's pretty simple. The card is charging about $30 a month in interest, and you're paying $60. Your payment covers your interest and makes a $30 bite into the balance.

Figure 3-1:
Paying off a
card at $60
per month.

	Making credit card payments: $60 payments on a $2000 balance		
		daily rate =	0.0005425
		# of days =	30
		factor =	0.016275
		payment =	60
Month	**Balance**	**Interest**	
	2000.00	32.55	
1	1972.55	32.10	
2	1944.65	31.65	
3	1916.30	31.19	
4	1887.49	30.72	
5	1858.21	30.24	
6	1828.45	29.76	
7	1798.21	29.27	
8	1767.48	28.77	
9	1736.25	28.26	
10	1704.51	27.74	
11	1672.25	27.22	
12	1639.47	26.68	

Making credit card payments: $60 payments on a $2000 balance

	daily rate =	0.0005425
	# of days =	30
	factor =	0.016275
	payment =	60

Month	Balance	Interest
	2000.00	32.55
1	1972.55	32.10
2	1944.65	31.65
3	1916.30	31.19
4	1887.49	30.72
5	1858.21	30.24
6	1828.45	29.76
7	1798.21	29.27
8	1767.48	28.77
9	1736.25	28.26
10	1704.51	27.74
11	1672.25	27.22
12	1639.47	26.68
13	1606.15	26.14
14	1572.29	25.59
15	1537.88	25.03
16	1502.91	24.46
17	1467.37	23.88
18	1431.25	23.29
19	1394.54	22.70
20	1357.24	22.09
21	1319.33	21.47
22	1280.80	20.85
23	1241.65	20.21
24	1201.86	19.56
25	1161.42	18.90
26	1120.32	18.23
27	1078.55	17.55
28	1036.10	16.86
29	992.96	16.16
30	949.12	15.45
31	904.57	14.72
32	859.29	13.98
33	813.27	13.24
34	766.51	12.47
35	718.98	11.70
36	670.68	10.92
37	621.60	10.12
38	571.72	9.30
39	521.02	8.48
40	469.50	7.64
41	417.14	6.79
42	363.93	5.92
43	309.85	5.04
44	254.89	4.15
45	199.04	3.24
46	142.28	2.32
47	84.60	1.38
48	25.98	0.42

Figure 3-2:
The end of
the $60
monthly
payments.

Your balance thus drifts down fairly slowly. How slowly? Well, here's the bottom of the table (see Figure 3-2). Four years later, you make your last payment of a mere $25 or so (the last bit of the last month's payment is too small for $60). You have now paid about $900 in interest, too.

Here's a little algorithm (*algorithm* is the proper mathematical name for a trick). Remember this the next time you see a camcorder that you just can't live without. I'm not saying that you can't have the camcorder (I wrote the revised version of this chapter on Christmas Eve!) — I'm saying that you should know what it costs. And, lest you worry that you're being lectured by some know-it-all who has never done anything impulsive, you may readily guess that my morbid interest in credit card math is the result of some fairly bruising personal experiences. Here's the trick:

If you buy something on a high-interest credit card and don't pay off the balance at the end of the month, the real price goes up 50 percent.

That is, if you buy something on a card with an interest rate of 18 percent or higher, and you make payments that are about double the minimum monthly payment, the purchase will have cost you about 50 percent more than the original price by the time the smoke has cleared. Let me be specific:

- ✔ The big-screen TV for $1,799 really cost $2,700.
- ✔ The $80 shoes cost $120.
- ✔ The $165 anniversary dinner cost $247.
- ✔ The $2,300 used motorcycle cost $3,450.
- ✔ The $200 jacket marked down to $140 (30 percent off!) actually cost $210.

The preceding rule isn't accurate to three decimal places, but it's surprisingly accurate, and it's certainly good enough to remember as a guideline.

Faster Payoff

Having given you that gloomy bit of calculation, I will now consider the cheerier prospect of a faster payoff. If you make payments of $120 per month, you'll be all paid up after 19 months and will have paid a mere $351 in interest.

At payments of $180 per month, you're out of the woods in a year, having paid out $223 in interest. The table in Figure 3-3 shows the financial results for different schedules of credit card payback, and the sidebar entitled "Interesting" explains how to do this sort of computation yourself. The take-home lesson, however, is pretty blunt and requires no further arithmetic.

Paying off $2000 at 1.67% per month interest		
Payment	**Months**	**Total of Payments**
$60.00	48	$2,905.96
$120.00	19	$2,351.09
$180.00	12	$2,223.30
$240.00	9	$2,166.67

Figure 3-3:
Paying faster.

Historically (meaning way back in the '70s and '80s), paying off balances slowly was almost a viable strategy. Inflation was roaring out of control, and you would presumably make the last payments in nearly worthless funny money. In the meantime, you would have had four years' worth of pay raises.

Well, I say "almost" because even back then it wasn't a brilliant strategy. Now, of course, it's a screaming disaster. The inflation rate is a few percent, the "real" interest rate (the difference between the inflation rate and the prevailing interest rates) is at an all-time high, and it's been years since middle-class Americans have seen serious raises. You make a purchase for $2,000, and at a slow payback rate, you have to come up with about $3,000. And you are no more likely to have the $3,000 later than the $2,000 you would have needed to make the purchase outright. You can dig yourself a pretty big hole this way, and the sides of hole are made of revoltingly slippery clay. By the way, this whole

Interesting

I did this computation in a computer spread-sheet for convenience, although you can certainly get the flavor of it with a simple calculator. You take the balance, apply the monthly interest rate, and then use the formula

> Balance = Old Balance + Interest − Payment

This method of finding out how long you make payments is pretty tiresome, however. If you have a financial calculator, you can do it this way. Remember that the elements of the standard payment calculation are payment, interest rate, present value (the same thing as "amount" in this context), and number of payments. If you enter the other three quantities, the calculator can solve for the number of payments. Probably the simplest way to figure how much you pay at a particular payment level is to enter the amount, payment, and interest rate per month and just let the calculator return the number of payments. Multiply the number of payments by the value of the payments you made, and you see what you've shelled out for the duration.

calculation assumes that you haven't put any other purchases on the credit card in the meantime. If you tend to go trolling at the malls and throw a few more purchases on the total every weekend, the situation is nowhere near as rosy as I portray it here.

Making the Minimum Payments

It may be the case that you've been through a lot, you're tough as an old boot, and you don't scare easily. That means you're ready for the numerical equivalent of the famous shower scene in *Psycho*. It's called "making the minimum monthly payments." Your credit card calls for a minimum payment of 2 percent of the balance, or payments of at least $10.

For all I know, you who are reading this book are taking some kind of blood pressure medication, are prone to anxiety attacks, or suffer from insomnia. For these and for other technical reasons, I will present the minimum payment schedule in two separate tables, "Starting Out" (see Figure 3-4) and "Finishing" (see Figure 3-5).

In the beginning

Sit down somewhere, get yourself a cup of tea, and take a look.

Figure 3-4 shows the first year of payments; it's pretty straightforward. You pay out the minimum monthlies, which total $382.70. After paying this amount, you have made nearly $81 worth of headway into the original loan amount. That's a little more than $300 worth of interest and $81 worth of payments on the balance, just to beat you over the head with the point. Pretty brutal, huh?

Near the end (?)

It probably has occurred to you that this payment rate against the balance gets you nowhere fast. So let's look down this giant calculation to, oh, 30 – 40 years in the future. Way down yonder in the table, *almost 40 years out,* things have settled down a bit (see Figure 3-5). Every month, you make a minimum payment of about $10. But the interest on your balance is about $8.50. At a few dollars a month, you're not exactly streaking for the end zone. But don't worry, pretty

465	104.88	1.71	10.00
466	96.59	1.57	10.00
467	88.16	1.43	10.00
468	79.59	1.30	10.00
469	70.89	1.15	10.00
470	62.04	1.01	10.00
471	53.05	0.86	10.00
472	43.91	0.71	10.00
473	34.62	0.56	10.00
474	25.18	0.41	10.00
475	15.59	0.25	10.00
476	5.84	0.10	10.00

Figure 3-4:
One year of minimum payments.

soon you'll be making the $10 minimum payments and catching up a little faster. In fact, 40 years after the purchase, a mere $9,089 in payments later, you open the billing envelope and see that you have only $5.61 left to pay.

If you pay at this minimum payment rate, this book will be a yellowed chunk of carbohydrate resembling a squashed and mummified roll of paper towels before you finish paying for your original purchase. Your *descendants* will have to sell your original purchase as an antique to settle the debt at this payment rate. That's why, in the fine gray print on the back of your credit agreement, careful legal provisions call for immediate payment in the event of your untimely demise.

Making credit card payments: $60 payments on a $2000 balance					
			daily rate =	0.0005425	
			# of days =	30	
			factor =	0.016275	
			payment =	60	
Month	**Balance**	**interest**			
465	104.88	1.71		10	
466	96.59	1.57		10	
467	88.16	1.43		10	
468	79.59	1.3		10	
469	70.89	1.15		10	
470	62.04	1.01		10	
471	53.05	0.86		10	
472	43.91	0.71		10	
473	34.62	0.56		10	
474	25.18	0.41		10	
475	15.59	0.25		10	
476	5.84	0.1		10	

Figure 3-5:
Minimum payments, much later.

Please explain!

What's going on here? It's really pretty simple. The situation resembles the nightmare credit arrangement described in Chapter 2, in which the payments are a few cents less than the total interest per month. In that case, the payments never make a dent in the principal.

In the wild and woolly days of 12 percent inflation, the situation was sufficiently unsettled that most cards were geared toward complete payback within three years, even though credit card companies were inflamed by the prospect of collecting huge amounts of interest from you. Currently, as inflation is under control and credit-card interest rates are still (in most cases) fairly stratospheric, the companies are perfectly happy to leave the debt on the books and collect interest. You, as a credit card borrower, are the best investment they've ever seen. If you had a spare thousand dollars, there's nowhere *you* could park it where you would earn a 15 or 18 percent return on your investment, unless you had a crystal ball and an exceptionally compliant stock broker. But the credit card issuers are singing "I've Got You, Babe!" and merrily allowing you the easiest payments ever seen, except that they go on for eternity.

Chapter 4

Mortgages

*M*ortgages are actually pretty straightforward financial deals compared to credit cards. That's because you're pretty much stuck with a barely changing loan balance for long periods of time even though interest rates may vary. On most credit cards these days, the interest rates are adjustable — the lender changes the rate according to a formula based on Federal Reserve Board rate decisions — and your credit card balance hops around as you make purchases. Checking your interest charges for a given month is thus an amazingly complicated job that is best left to a computer.

You can get two basic kinds of mortgages: *fixed-rate* and *adjustable-rate*. In a fixed-rate mortgage, you have a definite interest rate that doesn't change over the length of the loan, a definite time period in which to pay back the loan, and a fixed loan amount. That makes for a nice, clean calculation — you make a payment every month, and you pay the same amount until you close out the loan.

An adjustable-rate mortgage changes its interest rate according to the latest interest rate determination by the Federal Reserve Board (the Federal Reserve Board sets the rate it charges banks for funds, and the bank then charges you a somewhat higher rate). Interest rates usually change at some fixed interval, such as six months or a year. When the rate goes up, your payments go up. When the rate goes down, your payments go down. In an economic state where interest rates are just about guaranteed to change all over the place over ten years, adjustable-rate mortgages make a lot of banking sense.

The problem is that if rates go high enough, you can find yourself living in a house for which the payments are simply more than you can make. It's what you might call a sporting proposition — you are literally "betting the farm."

Doing the Fixed-Rate Mortgage Numbers

A standard form of mortgage these days is the 30-year fixed-rate mortgage. You can figure out the payments for this quite easily with a financial calculator, but for convenience I'll give you a little table here that lets you do the same calculation with just a four-function calculator (or even — gasp! — with paper and pencil). The author would like to thank D. Delfino for many stimulating discussions about mortgage problems.

This table gives the payments for a loan of $100,000 at a series of interest rates.

Table 4-1	Monthly Payments for a 30-Year Fixed-Rate Mortgage on $100,000
Interest Rate	*Monthly Payment*
6%	$600
7%	$665
8%	$733
9%	$805
10%	$878
11%	$952
12%	$1,029

Unlike the tables you might find in a real estate agent's loan handbook, notice that I don't keep track of pennies. Your real loan payment for the first entry would be $599.55. (It's sometimes easier for you to see what's going on if you don't have to look at so many digits.) Try a few examples.

A $100,000 loan at 11 percent

I just put this in to make sure that you were paying attention. Look it up in the table — it's $952 (actually, it's $952.32 if you want it to the penny).

A $79,000 loan at 7 percent

If you were borrowing $100,000, you could again get the number right out of the table — $665. Now, you can probably guess that if you were borrowing exactly half as much, namely $50,000, you might reasonably expect that the payments would be half as much, or $332 (and change). You would be right, too.

So the payment on $79,000 is likewise just in proportion. The payment will be

$$665 \times (79{,}000 \div 100{,}000) = 665 \times 0.79 = \$525$$

A $129,000 loan at 7 percent

The same rule applies for larger amounts, as you would also expect. To get a payment for $129,000, you find

$$665 \times (129{,}000 \div 100{,}000) = 665 \times 1.29 = \$858$$

A $153,450 loan at 9 percent

The principle is still the same, although the numbers look more complicated. The payment for $100,000 at 9 percent is $805, so you take

$$805 \times (153{,}450 \div 100{,}000) = 805 \times 1.5345 = \$1{,}235$$

A $100,000 loan at 8¹/₂ percent

This problem is a little different. The table doesn't give the payment for an $8^{1}/_{2}$ interest rate. But you might reasonably expect that the payment has to be somewhere between the payments at 8 percent and 9 percent. In fact, what else could it be?

The 8 percent payment is $733 and the 9 percent payment is $805. The difference between these values is just

$$\text{difference} = \$805 - 733 = \$72$$

Although it's not exact, you can get a payment that's correct within a few dollars just by taking half this difference and adding it to the 8 percent payment.

half the difference = $72 ÷ 2 = $36

8 percent payment + half the difference = $733 + 36 = $769

This process is called *interpolation* (in case you want the big word for it) and is illustrated in Figure 4-1. Just so you know, the exact on-the-money to-the-cent number for $8^{1}/_{2}$ percent is $768.91. This is why I'd rather show you a few tricks and a very simple table than print out every possible payment. It helps you understand the process, and, in fact, you could write this little table in permanent marker on the back of a $2 four-function calculator and declare it to be your mortgage financial calculator.

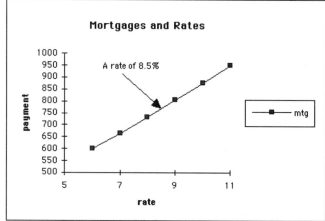

Figure 4-1: Finding other points in a table.

And now, a $139,500 loan at $8^{1}/_{4}$ percent

This is a sample real-world calculation because the amount financed is always some peculiar not-round number, and it seems that mortgages are usually offered in $^{1}/_{4}$ percent increments. Get out a piece of paper and take this in two steps:

1. **Determine the $100,000 payment at the fractional interest rate.**

   The rate of $8^1/_4$ percent is between 8 percent and 9 percent. Again, the 8 percent payment is $733 and the 9 percent payment is $805. The difference between these is just

   difference = $805 − 733 = $72

   Now take $^1/_4$ of this amount.

   $^1/_4$ the difference = $72 ÷ 4 = $18

   8 percent payment + one fourth the difference = $733 + $18 = $751

   The monthly payment on a $100,000 mortgage loan at $8^1/_4$ percent for 30 years is thus $751.

2. **Now adjust the payment for the fact that you're borrowing $139,500 instead of $100,000.**

   Instead of $751, the payment becomes

   751 × (139,500 ÷ 100,000) = 751 × 1.395 = $1,048

According to the financial calculator, the correct answer is $1,048.02. I'm intensely familiar with this number because a mortgage firm mails me a little ticket with this sum printed on it every month. That is a real deal by California standards — I just happened to get it because I moved to this adorable little town about nine months before prices blew sky high.

If you are curled up with this book somewhere in the Midwest and have just returned from shoveling snow or alternatively have had a picnic rained out, please console yourself with two facts:

- ✔ You at least can probably follow all the arguments in this book. Students in Iowa and Minnesota, for example, have the highest math scores in the world, beating out the students in all the countries of Europe and tying with Taiwan.

- ✔ In my little town, a two-bedroom wood frame house with a bad roof and foundation problems goes for about $300,000. No kidding. All we get to do with our money out here is make ridiculous house payments.

Shorter Mortgages

Thirty years is a long time. Thirty-year mortgages, in fact, were an invention of the late 1970s when interest rates got so out of hand that making payments on a middle-class income was practically impossible in areas of high real-estate inflation.

If you can handle the payments, a shorter mortgage is better. Think of it this way: You borrow a huge pile of money from the bank. You borrow it at an interest rate that's probably five points higher than what that same bank pays on a savings account. And you then start making your payments. But you're still making payments 30 years later. You have borrowed this money for a very long time. The interest has been compounding monthly through 360 compounding periods. Even if you haven't been following the numerical details of some of my arguments, you have to have a gut feeling that compounding interest 360 times is gonna cost ya!

So here's another table, showing monthly payments for both 30-year and 15-year fixed-rate mortgages.

Table 4-2	Longer versus Shorter Mortgages	
Interest Rate	*Monthly Payment for 30 Years*	*Monthly Payment for 15 Years*
6%	$600	$843
7%	$665	$898
8%	$733	$956
9%	$805	$1,014
10%	$878	$1,075
11%	$952	$1,137
12%	$1,029	$1,200

The payments at rates like 8 percent and 9 percent are about $200 per month more in the 15-year mortgage (compared to the 30-year version). That's $200 per month more for 15 years. Compared to the total payments on 30 years of the lower payments, the 15-year mortgage is a bargain.

The 9 percent comparison

Take the case of 9 percent as the interest rate. Here's the comparison:

- ✔ **30 years**

    Thirty years is 360 months, so you multiply the monthly payment by 360. That gives you

    Total = 805 × 360 = $289,800

- ✔ **15 years**

    Fifteen years is 180 months, so that's the factor now.

    Total = 1,014 × 180 = $182,520

The loan payoff total is different by more than $100,000. That's actual money that you have to produce. It's $100,000 *more* that you have to spend to buy the same house. This is serious business indeed.

The 12 percent comparison

At higher interest rates, the incentive to pay off faster becomes even more severe. For 12 percent, the figures become

- ✔ **30 years**

    Payments are $1,029, and 360 × 1,029 = $370,440

- ✔ **15 years**

    Payments are $1,200, and 180 × 1,200 = $216,000

The difference is now more than $160,000. Owing a bank lots of money for 30 years at 12 percent means that you pay nearly four times the loan value over the years.

Check out the graph in Figure 4-2. It makes an impressive case for trying to deal with larger payments on a shorter term.

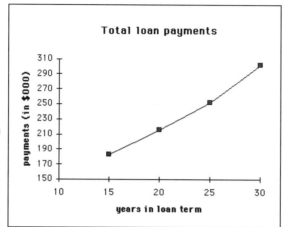

Figure 4-2:
Shorter
versus
longer
mortgages.

Adjustable-Rate Mortgages (ARMs)

Although you don't know what you will ultimately pay when you sign up for an adjustable-rate mortgage, here's a little test that you may want to try. Look at the real starting rate of the mortgage. This may take a bit of inspection of the paperwork because many of these have a trick *starter* rate that's not the real beginning rate. Sometimes the rate will be quoted at 2 percent or so below the real ARM rate as an inducement for you to take out the loan.

A typical 1995 deal might include a starter rate of 4.5 percent for the first 6 months and then jump to the real ARM rate of 6.5 percent. That increase will mean a bump up in your house payments of about $120 per month (on a 30-year term), which is probably survivable. But if you originally budgeted to survive a 6 percent mortgage and the rate over the years goes to 9 percent, you could have problems — after you get settled in, just have the family eat peanut butter and jelly for a whole week out of the month.

The big question is this: Will you get squeezed out altogether, finding yourself unable to make payments? The Federal Reserve Board is apparently terrified of the prospect that the American middle class will get back on its feet or, worse yet, that wages here might recover to levels that are commonplace in poorer European countries. Thus the Federal Reserve Board has shown an alarming willingness to raise interest rates four and five times in a single year. Look at the little table in Figure 4-1 or 4-2 depending on your mortgage term, and calculate what would happen to the family budget if the ARM floated up to 8.5 percent after a starting rate, in this example, of 6.5 percent. If you can't handle a two-point jump, you have serious prospects of being in trouble in a few years.

When you take out an adjustable-rate mortgage, make sure you understand 1) what the maximum the rate can be (there's a stated maximum) and 2) how fast the rate is allowed to change (every six months? every year?). If you have this information, you can at least figure what the maximum payment would be and when you would have to make it.

The Great Game and How to Win It

The medieval Christian church frowned on interest in general and felt that high interest rates (the kind that are typical now) were a sin. Similarly, Islam still forbids charging interest on deposits; in Islamic banking, your contract with the bank is phrased in terms of a joint venture, and your return is defined as your part of the profits from the joint venture. These religious scruples recognize that interest is a dangerous thing (try owing money to an illegal sports-betting operation if you don't believe me) and greatly amplifies the economic advantages of the *haves* versus the *have-nots.*

I realize that this is a math book, and I have tried (really I have) to keep my sermons to a minimum. But sometimes the numbers from simple math calculations just hit you over the head. Here are two examples based on the mortgage tables.

Fancy trucks and little houses

This particular numerical exercise is being played out daily in my own cozy home town. And if you replace the truck with a Jaguar, this exercise is also being played out in some of our finer suburbs.

Mr. Whattheheck has been offered a fine new job with an electrical contractor in the semirural town of Buendok, California. He gets to leave the big city and live in an area with interesting recreational possibilities, including hunting and fishing.

He can buy a new house with a loan amount of $109,300 and get an 8 percent fixed-rate mortgage. He also figures that now he should start driving a new pickup instead of his aging Ford Ranchero because of the fabulous recreational possibilities of the new area.

Here are his choices.

Plan A: Cheaper truck, shorter mortgage

Mr. Whattheheck gets a 15-year fixed-rate mortgage at 8 percent and a basic $12,500 truck with a 6.9 percent 5-year loan. From the data in the mortgage table in this chapter, you can determine that his monthly payments are as follows:

House payment = $956 \times 1.093 = \$1,045$

Using the table in Chapter 3, taking my word for it, or using a financial calculator, you can also find the truck payment:

Truck payment = $246

Payment total = $1,045 + 246 = \$1,291$

Plan B: Way-cool truck, longer mortgage

Mr. Whattheheck gets a 30-year mortgage instead and a totally awesome $24,250 truck (huge wheels, chrome transfer cases, roll-bar with halogen lamps, and killer interior trim package) with a 7.9 percent 5-year loan. Now his monthly payments (computed the same way as above) are as follows:

House payment = $733 \times 1.093 = \$801$

Truck payment = $490

Payment total = $801 + \$490 = \$1,291$

Now I'll look in on the two plans five years later. The truck is paid for in either case. In Plan A, a little work with a financial calculator shows that a balance of $86,092 remains on the home loan (in the first few years of all these loans, you're mostly paying interest). In Plan B, the home loan has a balance remaining of $103,911.

So the financial difference between these two plans is $17,819 after five years, minus the difference in the resale value of the trucks.

Here's the question arising from this case: If you are a regular old American middle-class person with a family income between $30,000 and $50,000 per year, *how long do you think it will take you to save $18,000?*

Could you save that much in five years? In this scheme, please remember, Mr. Whattheheck is making exactly the same payments every month. There's no particular strain on the finances to end up gobs of money ahead of the game.

What he has to forego is the pleasure of driving the bigger truck.

I am not preaching here in favor of small trucks, small TVs, refrigerators *without* ice makers in the door, or vacations near home. I just want to punch up the numbers so that you can see how this situation plays. You get to pick because you get to live with the results. Someday we'll all be dead, so maybe you should just have some fun now. But if you're not dead soon enough, all you'll have for retirement is pictures of the cool truck that you had 30 years ago.

Savings? What savings?

I will now suppose that you are a serious-minded character, immune to the blandishments of our overcooked consumer society. You don't have a 40-inch TV, and you don't replace the carpeting every two years.

Suppose you have a choice between taking out a 15-year fixed-rate mortgage at 9 percent or taking out a 30-year mortgage at 9 percent and putting $200 every month into a savings account at the same bank. Now you might see that borrowing money at 9 percent and lending it back to the bank is a curious proposition, but it's interesting to see the actual numbers.

When the 15-year mortgage has been paid off, you own the building. In this plan, you don't have a savings account, but you are sitting on piles of equity. If, at some time during the 15 years, you needed a cushion for emergencies, the bank would have been delighted to issue you an equity line of credit to use for emergency loans.

Consider the 30-year mortgage scenario. At the end of 15 years, you have a pile of savings in the bank and you have paid off some of the loan. The actual numbers (I'm assuming that this is a $100,000 loan) are

- ✔ Amount left on the loan: $79,330
- ✔ Amount in the bank: $49,202

For all your diligence in saving, you are worse off than if you had taken out the 15-year mortgage.

A simple rule

You have to look at your debts and see whether you have an alternative use for your money. I doubt you can find a reliable investment that yields 22 percent, so if you have a department store credit card at this rate, you don't have any better use for extra money than to pay off this card.

The way a bank works is simple: A bank takes in deposits, pays one rate of interest to depositors, and then lends it out at a higher rate to lenders. As a homeowner, you *may* sometimes find a use for your money that pays a higher return than you are paying to the bank. Remember you're one of the *bank's* investments, and your mortgage interest rate is the return it gets from *you*.

If you get a red-hot deal on a load of 500 knife sharpeners (10 cents apiece) and you can sell them at a swap meet for $5 each to gullible passers-by, by all means leave yourself $50 extra every month for this admirable activity. But recycling your money through the banking system works to the bank's advantage, not yours. It's always nice to have some cash on hand (at least it keeps you from putting impulse purchases on credit cards), but you should understand that if you owe a bank $100,000, you don't really have any *savings* per se — you have cash that you haven't used yet to cover your debts.

Chapter 5
Taxes and Paychecks

• •

In This Chapter

▶ Understanding what all those numbers on your paycheck mean

▶ Filling out your W-4

▶ Determining how much Federal tax to have withheld

▶ "Good enough" tax guessing

▶ Special advice for the self-employed

• •

*B*y the request of IDG Books Worldwide's president, this book has chapters on tipping, checkbook management, and puzzles. By the request of nearly everyone I know who writes for *Macworld,* my main publishing "home," it also includes this chapter on taxes.

Most magazines are put together by two kinds of people. One group consists of the hard-working, dedicated, in-house employees, people who actually show up in the offices every day and answer the phones and figure out what goes in the next issue. These people face the question: What do I put on my W-4 form so that my taxes come out right? They get married, they get divorced, they buy a house, they sell a house, and every time one of these events happens, the withholding amount on their paychecks should change. But how much change is right? That's one of the topics that I'll cover here.

The other group of magazine people consists of the writers and contributing editors, who as a rule show up nowhere on time, work mostly in fuzzy bathrobes at odd hours, and get paid for each assignment as a separate contract. When they hand in an article to an in-house editor, usually a week after deadline, the magazine authorizes a check to be mailed out. These people face the question: How much of this check do I set aside for taxes?

A Check Stub Decoded

Ladies and gentlemen, may I direct your attention to Exhibit A (see Figure 5-1), the centerpiece in our presentation of evidence of crimes against the taxpayer. Note that I have used the name John Doe, although I could equally well have used the name John Q. Taxpayer, the balding, mustached figure who appears in political cartoons, usually carrying heavy burdens or else having his pockets vacuumed. I generated this particular example with a standard payroll computer program, so the print is a little larger than it would be on an actual check stub.

Because this check report was generated in California, it includes more deductions than you might see on your own check. If you live in New York City, you would see even *more* deductions, as that great city needs lots of money to provide its inhabitants with the full spectrum of fabulous benefits pertaining to life in the Big Apple. If you live in Nevada, this check should make you glad that you live on the other side of the state line. I'll run through these items quickly so that I can get down to a discussion of how to deal with the Big Items.

Figure 5-1:
A genuine
fake pay
stub.

Emp #005		John Doe	
HOURS		**EARNINGS**	
REG1	3000.00	GROSS	3000.00
O/T1	0	FD-TAX	508.48
VAC	0	FICA	229.50
S/L	0	ST-TAX	128.28
		SDI	30.00
		NET PAY	2103.74

Here's the Magic Decoder Ring for this document:

- ✔ **REG1** is just the pay amount for the month. Poor Mr. Doe is evidently not collecting any O/T1 (overtime) or VAC (vacation) pay during this time period, but then again, he hasn't needed any S/L (sick leave). There are no bonuses or tips, as Mr. Doe is a plain old salaried employee.

- ✔ **FD-TAX** is the first big deduction item, Federal Income Tax withheld (it's 508.48) on the report. What's that as a percentage?

  To find what percentage of your income goes toward Federal taxes, divide the amount you pay in Federal taxes by the amount of your total pay . If you divide 508.48 by 3,000, for example, you get the number 0.169, so in percentage terms Mr. Doe is paying 16.9 percent in Federal taxes. Since Federal tax rates run from 15 to 31 percent (in 1995 anyway), he's near the low end of the scale.

- ✔ **FICA** is the tax for Social Security at a fixed rate of 7.65 percent. You can convince yourself that you are indeed paying that rate by multiplying

  $$3000.00 \times 0.0765 = 229.50$$

  which is the exact amount on the report.

- ✔ **ST-TAX:** Here in the Golden State, Mr. Doe finds himself turning over about 4.3 percent of his gross income to the State. Actually, he gets nicked much worse than that on sales tax, but sales tax isn't shown here. In ten states, you don't pay any State taxes, while only one or two other states are as voracious as California. But hey, we lead the *world* in new prison construction.

- ✔ **SDI** is California's state disability insurance, assessed at a uniform 1 percent across the board whether you work in a library or race motorcycles for a living.

Filling Out Your W-4: Deductions

As you probably know, there's not much you can do about two of these items, FICA and (if your state does it) SDI or the equivalent. These deductions are taken as a fixed percentage of your gross pay.

The Federal tax and State tax are computed by using information that you list on a form called a W-4. You are usually handed a copy of this form on your first day at a new job. You look at it for a minute, make some guesses, and hand it back to someone in the personnel department. Personnel enters your information into the payroll program, and the consequences appear on your paycheck.

The high-risk approach

You may decide to list on your W-4 that you are married and have 15 dependents. If Mr. Doe had done so, the payroll computer program would have decided that he was unlikely to pay much tax, because with this number of dependents he would have no taxable income left from a salary of $3,000 per month.

Over the course of the year, the program would take out almost nothing for either Federal or State tax (the computation of the two is linked). When he actually files the Federal and State returns, Mr. Doe's choices would be to 1) fork over about $7,000 that he won secretly in illegal sports betting, 2) work out a payment plan with the Feds, which involves truly unpleasant penalties, or 3) skip out to a country that doesn't extradite people for financial crimes, with Brazil being a favorite example from the movies.

In other words, overloading your number of dependents on this form is one of the most dangerous things that you can do with a pen.

The low-risk approach

You can avoid direct disaster by indicating on this form that you are single and have no dependents, whatever your actual situation is. If you are single and have no dependents, of course the amount withheld will be correct. If you have dependents or are married, the payroll program will take out more than is necessary, so you should get something back when you file your taxes. The only problem here is deciding whether you are happy in making an interest-free loan to the government.

If you habitually come up short at tax time, this approach may be the best idea anyway. Deduction computations are just math, but you may need to take human factors into account. If you are terminally irresponsible (that's not a slam — lots of my best friends are terminally irresponsible), you may be better off letting the Feds hold your money for you at the maximum tax rate, preventing you from squandering every nickel you get. Getting the taxes exactly right and investing the difference would make more fiscal sense, but if you weren't going to invest anyway, figuring your W-4 to perfection doesn't make practical sense.

Getting It Right

There are two ways to figure out how much you should be paying in withholding in more complicated situations. If you are married and have a few dependents but are buying a house in one of the many overpriced neighborhoods on the East or West Coast, you will find that the interest part of your mortgage payment has a bigger impact on your taxes than the dependents do.

Figuring out how to adjust the withholding while taking into account giant mortgage payments is a matter of several thousand dollars one way or another, so it's worth considering for a few minutes. The back of the W-4 contains some advice for doing the computation, couched in the uniquely unhelpful language that Federal form designers have made a traditional specialty. If you can follow all the computations on the back of the W-4 form, you can skip the rest of this chapter. In talking to a half-dozen income tax preparers, I have been told that this is a big source of confusion. The aim here is to give you some guidelines that at least won't get you into trouble.

The math way

Here's one approach to figuring the number of deductions. Please observe that this is only a sample, a suggestion, a hint, a back-of-the-envelope crude estimate. Tax practice can and does change at any time. I just want to show you a little calculation here. Your actual mileage may vary.

1. **Take the gross salary of the "tax unit."**

   If it's just you, it's just you; but if you're married, then it's both of you. I'll use $75,000 so that you have some real numbers to follow.

2. **Subtract the interest part of your home loan from the gross salary.**

   Your mortgage holder sends you a statement every year that gives you this number, which changes a bit from year to year. In this case, I'm going to use the figure $14,000, which is about right for payments of $1,300 a month or so at five years into the mortgage. That brings the taxable income down to

   $75,000 − $14,000 = $61,000

3. **Subtract $2,450 for each of your dependents.**

   That's one if it's just you, two for a married couple, four for Mom, Dad, and two kids, and so forth. Just listing real dependents is simpler, but it has a bias in favor of higher withholding. You *want* some bias in that direction so you don't get into trouble. In this case, I'll figure it for Mr. and Mrs. Taxpayer and their children Deirdre and Fang. That takes the taxable income down to

$$\$61{,}000 - (4 \times \$2{,}450) = \$51{,}200$$

4. **Multiply the number that you got in step 3 by the applicable Federal tax rate. You can look this up in last year's return.**

The Federal tax rates range from 15 percent up to 31 percent at the moment, and this level of income is taxed 28 percent. That means, without thinking through any other deductions, that you pay the following amount in Federal taxes annually:

$$\$51{,}000 \times 0.28 = \$14{,}280$$

Per month, that's

$$\$14{,}280 \div 12 \text{ months} = \$1{,}190$$

Mr. and Mrs. Taxpayer should check their pay stubs and make sure that their Federal withholding amounts from the two stubs add up to something in the neighborhood of $1,190 per month. Actually, this withholding amount is a somewhat generous overestimate with a built-in margin of safety.

The easy way

As a single person, you don't have much withholding risk. In returns for married people filing jointly, the risk is twofold:

- ✔ You both claim the same number of dependents on your W-4, which means that you are both claiming each other and the kids.

- ✔ Your two incomes added together put you at a higher tax rate than either of you would face individually. If one of you makes $65,000 and the other makes $25,000, that $25,000, which by itself would be in a modest tax bracket, is now taxed at the highest rate.

Here's one way to correct against the risk of being underdeducted. If you have large mortgage payments in interest to use as a deduction, take the annual amount and divide it by $2,500 (it's easier than using the exact figure, $2,450). Doing so converts your house into a number of dependents. Round the number *down* — if you get 3.4, for example, use 3 as the number. Now, whichever one of you has the higher income should fill out the W-4 stating as the number of dependents the number of actual live warm bodies plus the number from the house. The person with the lower income in the pair then fills out his or her W-4 form listing *no* dependents. This method gets the total withholding right within a few percent.

Self-Employment and Taxes

One other tax consideration needs to be mentioned. In the fast-paced, downsizing world of business in the 1990s, many people who used to be managers inside a company find themselves out the door, working on contracts as consultants. Writers and other self-employed people have adjusted to the tax differences between the inside and outside situations, but they may be new to you, so here are some notes.

The tax circumstances are now a bit different. On the inside, 7.65 percent of your salary was removed for FICA. Your employer was also paying the exact same amount, 7.65 percent of your gross pay, as an employer's contribution to FICA. On the outside, you get to pay your 7.65 percent *and* the employer's 7.65 percent, too. In other words, you now pay $2 \times 7.65\% = 15.3\%$ into FICA.

This tax is taken on your whole earnings, not the amount left after other taxes. If your net income, after expenses and deductions, is $50,000, your FICA contribution is

$$\$50,000 \times 15.3\% = \$7,515$$

Because you're at the 28 percent tax rate at this income, your Federal income tax is

$$\$50,000 \times 28\% = \$14,000$$

Put another way, every time someone sends you a consulting payment for $2,000, you are supposed to earmark $860.60 for Federal government payments alone. If you live in a state with a reasonably peppy income tax rate, it means that you get to keep a little less than half of what your checks say that you earned.

Someday, you may find yourself sitting in a boring meeting as an employee and thinking about doing something more entrepreneurial than taking orders. You should figure that, to compensate for this FICA shift and the loss of various employee benefits, you will need to take in nearly 20 percent more gross income as an outsider to match your income as an employee.

Some Stern Words of Advice

These methods of figuring approximately correct withholding are just that: *approximations* suggested as general-purpose guidelines by various tax authorities. But because this is a math book, let me tell you something that I can prove with an accurate statistical sampling:

- ✔ You can get math puzzles in the paper wrong and not suffer for it at all.
- ✔ You can foul up your checkbook by a small amount and suffer a moderate amount.
- ✔ If you get your taxes wrong, you've got *real* trouble.

Because you are reading a book called *Everyday Math For Dummies,* you may want some help with math questions for which wrong answers are expensive.

One tack to take might be to poll your friends, relatives, and coworkers to find a tax accountant. If you have enough deductions to itemize — and owning a house usually means that you are in this category — you will find that even a moderately competent professional tax preparer can save you more than he or she costs in fees.

I should also tell you that although many excellent tax preparation software programs are available, they are just no substitute for a good preparer. Someone who does hundreds of returns a year and has many years of experience in dealing with the IRS is almost certainly going to get better results than you will get on your first outing with a new computer program.

Chapter 6
Simple Investment Math

• •

In This Chapter

▶ Basic, safe stuff: Savings-related deposits

▶ The stock market

▶ Some numbers for stocks

▶ Funds of various kinds

• •

*T*here are two ways to end up with a reasonable amount of money. One way is to get a high-paying job that you can do forever and then work at it as much as you can. The other way is to collect some money and let the money work for you.

The money never gets tired. It works 24 hours a day. It takes no vacations. Because the tax laws of every country are written for the benefit of the people with big deposits rather than small loans, much of the money that your money earns gets a tax break. The bottom line is that your prospects of retiring someday are usually based on getting some money and putting it to work rather than working harder yourself. Again, go buy a copy of Eric Tyson's best-selling *Personal Finance For Dummies,* which has lots of advice and phone numbers and comparative investment plans.

Safety First

The simplest type of investment is any type of savings-related deposit with a fixed yield. This type has the lowest yield of any possible investment because it carries the lowest risk. In a year when you can make about 3 percent interest in an ordinary savings account, you might make 45 percent on your money by investing in just the right high-tech video-game stock. There are two kinds of fixed-yield investment to consider here, discussed in the following sections.

Negative numbers: credit cards

The first of these "investments" to consider is the one you may be making right now to make someone else rich: credit cards. You don't have any savings until you have paid off your credit cards. You also have to decide whether you can find an investment that pays a higher interest rate than the rate you are paying on your home loan. For starters, look at the interest rates of your loans and compare them to the interest rates you can earn with your money in different deposits.

You may have an occasional opportunity to make a big score on borrowed money. Most of the time, however, it's not true, and it's a dangerous approach to take.

Positive numbers: CDs

Bank certificates of deposit (CDs) are insured by the federal government. You not only get your money back, but you also get the interest that you were promised. CDs as a rule pay much better interest than ordinary savings accounts, but they tend to pay a bit less than Treasury bills.

It's difficult to tell when you should try to lock in higher rates by picking a longer-term CD (five years) instead of a shorter-term one (one year or less). Sometimes, if rates subsequently go up in the longer term, what you have done is to lock in the lower rate. Rather than try to figure this out, you may want to scan the popular finance magazines (*Money, Kiplinger's, Worth*, and so on) and find a bank that's paying a half-point higher interest than the banks nearest you. Not all banks pay the same rate, for mysterious reasons of their own. The simplest CD strategy is just to find the highest-paying CD available to you.

Stocks

I'm only discussing investments here because the stock market itself gives rise to many mathematical questions. Here's one of the fundamental queries:

Can you figure out a way to prove whether someone else has figured out a foolproof system for selecting stocks?

The best evidence so far is that no such system exists. The evidence shows that no individual — or computer program, these days — has collected all the money in North America. If it were possible to predict the price histories of individual stocks, the person doing the predicting could collect any amount of money at will. This clearly does not reflect what has actually happened.

Nonetheless, while individual stock prices bounce all over the place in time (see Figure 6-1) based on the sometimes random events that occur in the business of individual companies, the stock market taken as a whole has been predictable, at least on long-enough time scales. As stock brokers never tire of pointing out, someone who was holding a big collection of stock at the time of the Great Crash of 1929 would have actually shown a profit by hanging on to the collection for six or seven years. Investors who held on through the crash in the late 1980s recovered within a year or so!

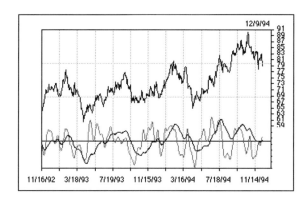

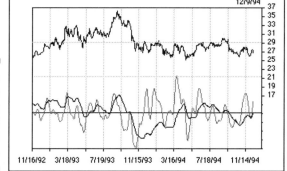

Figure 6-1:
The nearly random behavior of individual stocks.

The reason you can expect a steady increase, over decades, in stock prices is that the stocks are an index of an expanding economy and tend to adjust for the increasing valuation of the net worth of companies and for inflation as well. The whole country increases in value, and the stock market increases in value. As long as the economy shows some net growth, it's reasonable to expect stock prices to increase. Studebaker may disappear and Microsoft may pop on the scene, but the aggregate keeps growing.

Characterizing individual stocks

The lure is simply that if you could correctly predict the direction of individual stocks, you could make an immediate fortune. Therefore, despite wonderful statistical evidence that this can't really be done, at least not for day-to-day movements in price, the level of research interest in this matter is *intense*. There are at least three useful numbers for characterizing the behavior of individual stocks, and if you take them as starting points for meditation rather than as scientific predictors, it can't hurt to understand how they're computed.

As a way of considering how much growth in your investment you will see over different time periods for different rates of return on your money, consider how long it takes money to double. Suppose you have a 20-year investment horizon. In 20 years, you'd see the following results:

- ✔ 6.5% annual return produces 3.52 times the original investment.
- ✔ 7.8% annual return produces 4.49 times the original investment.
- ✔ 8.6% annual return produces 5.21 times the original investment.

Scrounging around year after year for a few percentage points' better return than mere bank CD interest can mean the difference between retiring with an adequate nest egg and just squeaking by in your sunset years.

Price/earnings ratio

A line in the stock market listing in the newspaper looks something like this (each entry has little variations):

Stock	Div.	P/E	Last	Chg.	YTD
XYZCor	1.04	16	$65^{1}/_{8}$	$+ ^{7}/_{8}$	+11.3

This is straightforward enough. *Div.* means *dividend*. This line says that if you bought a share of this stock, the company would send you a check for $1.04 once per quarter. Actually, for the stock at this price, that's the equivalent of a 6.4 percent annual interest, assuming that the dividends stay the same. *Last* is just the closing price, quoted in eighths of a dollar for a bit of antique charm (it's just amazing that the U.S. didn't retain pounds, shillings, and pence — seems like a natural).

Chg. is the change since yesterday, so the price of this stock has moved up by $0.875 since the day before. *YTD,* a relatively new item in stock reports, stands for *year to date.* It's the change in the stock price since the beginning of the year. In this case, the stock is up over 11 percent for the year, so the investors are happy.

The number that contains a bit of extra interest is *P/E,* which stands for *price-earnings ratio.* Just as it says, it's the price of the stock divided by the earnings per share. Different industries tend to have different characteristic P/E ratios, but within this distinction, lower P/E ratios (5 - 9) indicate a stock with strong earnings compared to its share price, and higher ratios (over 30 or so) indicate a stock whose price reflects investor optimism rather than actual earnings performance. Some investors simply collect stocks for which the P/E ratio is lower than the industry average; in fact, this method would probably work except that everyone knows it, which tends to reduce its effectiveness.

Variability and volatility

Variability is an index that shows the spread in the prices of a stock for a single year. Figure 6-2 shows a high-variability stock contrasted with a low-variability stock. With a low-variability stock, you have the assurance that the price will stay pretty stable, combined with the sobering assurance that stability means limited potential for gain.

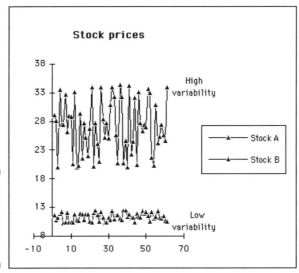

Figure 6-2: Low- and high-variability stocks.

You are usually obliged to call a broker to get a number for the variability of a particular stock. One of the uses for this number is to suggest whether the stock is a subject of much interest among speculators — a low variability means that the stock is not getting pushed around much by the big boys. You need to know it because low-variability stocks in established industries are often relatively safe investments.

Volatility is another fancy number, computed data on both the whole market and the individual stock. It measures the tendency of the stock to move along with the direction of the market. Because you don't have most of the numbers that you need to calculate this figure, you get a particular stock's volatility number (called *beta* by the hipsters) by asking a broker.

If a stock has a beta of 2.1, then the stock is a peppy little number whose price moves up (or down) a bit more than twice as fast as the rest of the market. If it has a beta of 0.3, it moves only 30 percent as fast as the other stocks in aggregate. A stock with a volatility of 0.0 never changes in price, no matter what the market does.

Characterizing the Whole Pile

Because individual stocks carry considerable risk, investors and investment companies seek safety by buying stocks in groups. Rather than take a chance on picking one stock alone, where the company, for example, could get hit with a lawsuit and start hemorrhaging money, investors buy collections of stocks on the assumption that they can't *all* be dogs. When you invest in many stocks at the same time rather than one, you reduce your investment risk. That's why there are different sorts of *funds*, or collections of stocks. Here are just a few.

Index funds

Investment group stock packages are usually called *funds,* although there are also funds of other investments (Treasury bills, for example). One of the simplest funds, although it is a relatively recent invention, is the *index fund.* The managers of these funds simply assemble a portfolio of the stocks that make up one frequently reported index, the Standard & Poor 500 index. If you buy shares in such a fund and hear on the radio that "the market" was up or down, it means that *you* were up or down. As an idea, this fund is as simple as it gets.

One of the advantages of this kind of fund as compared to standard *mutual funds,* which are carefully selected portfolios of stocks, is that nobody has to do any selecting. Instead of rooms full of analysts sifting through hot tips day and night, the managers buy the same stocks that make up the index. This is just a collection of prominent, large companies, so you have some additional financial security. If the stock market collapses, you collapse, too, but historically, this bet is pretty safe.

Mutual funds

Naturally, it stands to reason that very clever people, studying carefully the conditions in individual industries, should be able to assemble a portfolio of stocks that outperforms some randomly chosen average. Lots of things "stand to reason," but this proposition is a little trickier to establish. I'll do an experiment.

Ups and downs

Figure 6-3 shows the results of my little experiment. I have made up little funds of different stocks, selected randomly from a period of a few months when the overall market wasn't particularly gaining or losing. I took a sample from this market because fund results are just compared to overall market performance anyway. You see five portfolios of six stocks each. Individually, the stocks were all just going up or down during the time period of the test.

Starting price	Portfolio 1	Portfolio 2	Portfolio 3	Portfolio 4	Portfolio 5
	Final price of stocks in portfolio				
$20.00	$16.25	$19.38	$22.65	$17.96	$17.30
$30.00	$29.88	$31.37	$31.15	$30.79	$34.69
$40.00	$46.13	$47.03	$37.69	$42.24	$37.84
$50.00	$41.00	$46.88	$54.11	$56.70	$52.40
$60.00	$66.29	$55.83	$71.60	$61.30	$69.82
$70.00	$81.46	$64.77	$67.20	$67.67	$71.79
$270.00	$281.00	$265.26	$284.41	$276.65	$283.85
Percent change	3.91%	-1.79%	5.07%	2.40%	4.88%

Figure 6-3: Random funds.

The question to consider is this:

What does it mean if a fund does better than the overall market in the short run?

Remember, there is no manager for any of these funds — the numbers were generated at random by a computer program. In fancier, more complicated models, you can simulate the performance of real mutual funds in the real market. What does it mean that this random fund selector mimics the real world so well?

The answer appears to be that few funds have ever been able to perform better than the market on a very long term (decades). If you are interested in this (and the question of how to pick a fund in general), please consult *One Up On Wall Street* (Penguin, 1989) by Peter Lynch, generally acknowledged to be one of the best fund managers of all time.

The simian system

This question has been answered by Joleen, the orangutan at Marine World/ Africa USA. The business section of the *San Francisco Chronicle* sponsored a portfolio selection contest about a year ago that included nine professional fund managers with good reputations and the ape.

The result? *Joleen came in fourth out of ten in the competition.*

Draw your own conclusions. Actually, some advantages of having an orangutan as a fund manager would be low overhead (bananas, for example) and less tendency to change the portfolio mix. For that matter, an orangutan as a broker would probably never call you with "hot tips" that simply represent the broker-age firm unloading its own purchase of some dog stock onto you!

There's a "scientific" explanation for Joleen's relative success. Under one theory about market behavior, you simply can't get your hands on any information that hasn't already been factored into the stock price. It doesn't matter that Joleen has no telephone — she couldn't get any worthwhile tips anyway. One bit of support for this theory is the historical perspective that only really illegal insider trading of the most scandalous kind produces consistent, guaranteed better-than-the-market returns.

Part II
Business Math

In this part...

Officially, this part deals with business math, but it's really more about how *you* interact with businesses. It begins the theme of interpreting the swirl of numbers presented to you every day (you'll see more later). You can get a pretty accurate picture of reality from information in magazines, newspapers, and even TV news, but much of the time you have to preprocess it with a bit of math to make sense of it.

Chapter 7

Percentages

• •

In This Chapter

▶ Percentages and fractions

▶ Mark-ups and mark-downs

▶ Useful percent tricks

▶ Percent idiocy in the news

• •

*P*ercentages, I'm afraid, are always around. They make a convenient way to concoct some fake drama from ho-hum statistics, and they make it possible to make deals at a sale seem better than they are. Traditionally, interest rates are expressed as percentages, introducing one more little bit of confusion into calculations with financial calculators. The numbers themselves can tell stories quite eloquently, but percentages sometimes rear their ugly heads when there's incentive to make the story bigger, smaller, or fuzzier.

Everyone could have agreed, long ago, that all references to numbers would use just the numbers themselves. Here's how that would work:

✔ You go into a clothing store and they're having a sweater sale. The sign would say, "$39.99 sweaters now cost $19.99. Don't worry about what it says on the tag — our cash registers are programmed to get it right."

✔ At the end of the year, the police chief of Nothingville reports that there were three home robberies in the last year instead of two. The *Nothingville Times-Dispatch* could then headline, "Three Home Break-Ins Last Year."

✔ You read a psychology report in the newspaper. It states that in a sample of 20 people tested, 10 coffee drinkers and 10 non-coffee-drinkers, 6 of the coffee-drinkers chewed their pencil erasers while only 5 of the non-coffee-drinkers did.

These little examples contain all the information you need in the numbers of their respective stories. Now read the equivalent statements in percentages and decide for yourself whether the statements become any clearer:

- ✔ "Sweaters Now 50% Off — Discount Taken at the Register"
- ✔ "Home Break-Ins Up 50% in Nothingville"
- ✔ "Coffee Held Responsible for 17% Increase in Eraser-Chewing"

Percentages and Fractions

A percentage is just a fraction. But it's a fraction expressed as a fractional part of the number 100 — it's from a Latin term *per centum* that means "per 100." Explanations of percents result in some major weirdness in math books. Please let me explain what I mean by *weird* in this context.

Practically every math book on the market offers as a tip or trick that you can find 25 percent of some quantity just by dividing that quantity by four. For example, 25 percent of the number 32 is just 8 because $^{32}/_4$ is 8.

This isn't a tip.

This isn't a trick.

Twenty-five percent *is* one-fourth.

Twenty-five percent is just a slightly more confusing way of expressing the number one-fourth. The ancients, who had no calculators, had a distinct preference for expressing fractions in the form ($^1/_{number}$) because they used fractions as directions for dividing actual physical objects (grain, bricks, or buckets of olives, for example). If you have a pile of bricks, it's pretty easy to make $^1/_3$ of the pile by arranging the bricks in three lines. If you want to find 33 percent of a pile of bricks, you have to convert the percentage to a fraction anyway.

Table7-1 shows common fractions and their equivalent percentages. Realistically, in a world of $2 calculators, you need to know only a few of these numbers. You should know that 50 percent is the same thing as one-half — hey, you probably *do* know that 50 percent is one-half. But for the rest of the percentage problems in this chapter, I recommend that you do whatever it takes to get the right answer (I'll tell you how), and more often than not, that means converting each question to a foolproof calculator problem.

Table 7-1	Values Expressed as Fractions and as Percentages	
Fraction	**Percent**	**Decimal Fraction**
$^1/_{100}$	1%	0.01
$^1/_8$	12.5%	0.125
$^1/_4$	25%	0.25
$^1/_3$	33.3%	0.333
$^1/_2$	50%	0.50
$^2/_3$	66.7%	0.667
$^3/_4$	75%	0.75
$^5/_6$	83.3%	0.833

If you have to remember one fast rule (again, it's a definition, not a trick), remember this:

To convert a percentage to a decimal fraction, move the decimal point two places to the left. Thus

18% = 0.18

37.5% = 0.375

100% = 1.00

Shopping

Enough of this abstract palaver. This book hopes to keep your attention by focusing on money. Let's go to the store and decode some common situations.

Everything 20 percent off

When everything in a store is marked down by 20 percent, it means two things, both of which are the same, mathematically speaking:

✔ Everything has been marked down by $^1/_5$ of the original price.

✔ Everything now costs $^4/_5$ as much as it did before.

Now try some arithmetic. You go to an auto parts store, where a small socket-wrench set cost $17.99 before the sale. Now it's 20 percent off. Twenty percent is not just $^1/_5$; by definition, it's 0.20, expressed using a decimal point.

First, multiply 17.99 by 0.20.

$$17.99 \times 0.20 = 3.598$$

What does 3.598 mean? It means $3.60, because stores have no way to charge you $^8/_{10}$ of a cent. So the new price is

$$17.99 - 3.60 = 14.39$$

Now, just to illustrate the second statement, note that all of something is 100 percent. If you take away 20 percent, what's left is 80 percent. This is the same as saying that $^4/_5$ is left when you take away $^1/_5$ (see Table 7-2). By using this convenient one-step method, you get a new price of

$$17.99 \times 0.80 = 14.392$$

Again, there aren't fractions of cents, at least in stores, so the price is $14.39.

It's kind of odd, really, that it never became the custom to say, "Everything in the store is now 80 percent of the old price." I guess the idea is that the sign should bring to mind the lump of money you keep, not the lump you pay.

Table 7-2	Fractions and Percentages in the Store		
Original Price	*Discount %*	*Discount in $*	*Final Price*
$17.99	30%	$5.40	$12.59
$24.99	35%	$8.75	$16.24
$100.00	20%	$20.00	$80.00
$219.00	15%	$32.85	$186.15
$499.00	18%	$89.82	$409.18
$7.99	25%	$2.00	$5.99
$139.99	50%	$70.00	$70.00
$59.99	20%	$12.00	$47.99
$199.99	25%	$50.00	$149.99

Additional 25 percent discount taken at the register

Suppose you're in a department store looking at a jacket that started out at $79.99. The little computer-printed tag now has a line drawn through the $79.99 in red ball-point ink, and the price on the tag says $39.99. That means that the jacket has already been discounted by 50 percent.

Can you see why? Follow these steps:

1. **Round the original price to the nearest $10.**

   Think of $79.99 as $80, since, after all, that's about what it is.

2. **Round the sale price to the nearest $10.**

   $39.99, similarly, is close enough to $40.

3. **Compare the original price to the sale price.**

   The price went from $80 down to $40. In other words, it got cut in half.

   The fraction $1/2$ and the percentage 50 percent are the same thing. Cutting the price in half means a discount of 50 percent.

But now you get an additional 25 percent off the marked price when you actually buy the jacket. What does this amount to? Figure it out in one of these two ways:

- ✔ One way to look at it is that you get a discount of 0.25 times the new price of $39.99. You can do this calculation pretty easily in your head if you recognize that $39.99 is $40 and 25 percent off means $1/4$ off. One-fourth of $40 is just $10, so the new price is going to be $30 (although in store arithmetic, that undoubtedly means that the new discounted price is $29.99).

- ✔ The other way to look at it is that 25 percent off means that 75 percent of the price is still there.

   $39.99 \times 0.75 = 29.99$

You get the exact same thing, as you would expect.

Mark-Up

Stores, of course, buy all their merchandise at a discount from the price you see. In grocery stores, this *mark-up,* as it's called, is relatively modest. In jewelry stores, it's relatively outrageous, sometimes as high as 300 percent.

What's the meaning of a 300 percent mark-up? The store buys a ring at a price of $80. A 100 percent mark-up would be $80 added to the original $80, giving the ring a shelf price of $160. A 200 percent mark-up would be twice as much, or 2 times $80 added to the original $80, for a price of $240. And a 300 percent mark-up means a final price of $320.

Some of this is a bit confusing, which is often the case with percentages. In this case, it is actually simpler to say that merchandise gets marked up by a factor of four instead of referring to the same operation as 300 percent mark-up. A formula for this mark-up would be

final price = [1 + (percent mark-up/100)] × original price

Plugging in the numbers for this jewelry example, you get

final price = [1 + (300/100)] × 80

= (1 + 3) × 80

= 4 × 80 = 320

Of course, this is one of the reasons that some jewelry stores post signs saying, "Pre-Christmas Sale! Pearls and Diamonds 50 Percent Off!" If the store marks up the merchandise by a factor of four when it first comes in the door, it can mark it back down by a factor of two and still make money.

Calculators and Percent Keys

Amazingly, there isn't much consistency in the way you use the % key on different calculators. Calculators from Texas Instruments, for example, use one style. If you want to find a 25 percent discount on $65, you press this sequence:

65

−

25

%

=

and you get the answer 48.75. I like this style because it follows the way you would state the problem out loud. As an unsolicited plug, I think that no home should be without a Texas Instruments BA-35, one of the least expensive financial calculators.

Alternatively, most Casio calculators do things a little differently. On a Casio, you do the same problem by pressing these keys:

65

×

25

%

−

This order seems a little strange to me, but it works. Of course, the idea that you would actually have to read the instruction booklet that came with your calculator is simply terrifying, but sometimes you just have to bite the bullet.

Ups and Downs

One odd feature of doing calculations in terms of percentages is that a change described as "down 25 percent" followed by a change of "up 25 percent" doesn't get you back to the original number.

Suppose there were 150 fatal traffic accidents in your county last year. (I hope this is because it's a big county and not because it's particularly dangerous.) Now, what's the number of traffic fatalities if the number goes down 20 percent?

150

$- (20\%$ of 150, which is $0.20 \times 150 = 30)$

$= 120$

So a 20 percent drop in traffic fatalities gets you to 120 from 150.

What happens if, in the following year, the number of fatalities goes up by 20 percent? Remember that now you're starting with the figure 120 as the basis of the comparison. You get

120

$+ (20\%$ of 120, which is $0.20 \times 120 = 24)$

$= 144$

Ups and Downs in the News

This business about mark-up is closely related to one of the main uses of percentages, the daily announcement of percentage changes in some quantity or other.

Here's an example of how the calculation works. Last year, 15 cars were reported stolen in Burgertown, a city of 100,000 people. In the preceding year, 12 cars were reported stolen. There are two ways to calculate the percentage change in car thefts:

> ✔ **Using a difference**
>
> The net change in car thefts is $15 - 12 = 3$, and the base for the change is a first-year figure of 12. So the change is
>
> $$\text{change} = 3 \div 12 = 0.25 = 25\%$$

✔ **Straight calculator work**

Divide the number for the new year by the number for the preceding year. That's 15 divided by 12, or

$$15 \div 12 = 1.25$$

Last year was 100 percent, equivalent to 1. To compare the two years, subtract last year (1.25 − 1 = 0.25) and keep the 0.25. That's 25 percent. You can look at the number 1.25 and say that this year's thefts are 125 percent of last year, or you can say that this year's thefts are up 25 percent from last year.

The problem with this style of reporting as compared to having the original information is that you become accustomed to thinking of *changes* as significant, whereas the actual *numbers* are significant. In this case, a change from 15 to 12 reported car thefts in a town of this size really doesn't represent a real trend — that size of fluctuation would be expected. Also, this report doesn't point out the main fact, which is that car theft is not a big problem for this community.

All numbers in reported statistics fluctuate by a certain amount. One good estimate of the natural amount of variation from one sampling period to the next is

$$\text{variation} = \sqrt{\text{quantity}}$$

If an area reports 10 car thefts per year, it would be more or less expected that this number might naturally bounce up or down by as much as 3 cars from one year to the next. If your area records about 20 embezzlements per year, you can expect the number not to be exactly 20 every year but 20 plus or minus 4 or 5.

In other words, 18 embezzlements one year and 21 the next doesn't mean that an accounting crime wave has started. Rather, it just means that the number will change from year to year, sometimes increasing and sometimes decreasing, by a predictable amount. If, in this case, you get increases of more than five cases two years in a row, *then* it's time to see whether something strange is happening.

I discuss this whole problem at length in Chapter 19, a little study of how to interpret numbers in the news. Part of the problem is that they're going to print that newspaper every day, whether there's really anything to report or not. That's why sports, weather, and the stock market are such prized news segments — they're guaranteed to provide some sort of fluctuating numbers no matter what.

Chapter 8

Sales and Trends

*Y*ou can't get through a whole day without reading or hearing a prediction of a trend. Okay, I take that back. If you're lost backpacking in the Sierras without a radio, you won't hear about trends. Instead, you'll be projecting your own: "At the rate I'm using up my supplies and the rate at which the weather is getting worse, I'm in big trouble in three days." Even if you're in a Zen monastery on the California coast, isolated from the rest of the world, little internal predictions run rampant: "If the last few days are any guide, I'm not actually going to be able to get out of this lotus position by myself after one more hour."

The point of these remarks is that the tendency to predict the future from the past seems to be built into the human nervous system. The daily bombardment of such trend predictions as the following is simply the media-amplified outcome of the natural human predilection for personal forecasting:

- ✔ Retail sales will be up 2 percent this Christmas season.
- ✔ The Chicago Cubs will blow it this year and every year until the sun burns out and becomes a brown dwarf star.
- ✔ This could be the wettest winter on record in Oregon.
- ✔ Trade with China will increase by 30 percent over the next decade.

That's why I've included a chapter on forecasting in *Everyday Math For Dummies.* Forecasting is indeed an everyday matter, and it uses math. The big problem for both the consumers and producers of forecasts is that the math involved is often plain wrong or is a misapplication of a technique that would be valid in some other context. And because the great driving statistical force in business is the quest to predict sales from one quarter of the year to the next, this chapter will discuss in some detail the thorny problem of sales prediction.

Looking at Trends

Not all trends are created equal. Some have a real underlying basis, and some are just random fluctuation. For example, the trend for the population of the United States to increase in Florida and to decrease in North Dakota has been going on for 20 years, and people in both states can give you meteorological, social, and economic reasons for it. It's still cold in North Dakota, too, so this trend may continue.

On the other hand, you can often hear business news commentators remark in grave tones, "Well, the market has been down for three days in a row now," when the total decline in the Dow-Jones Index over the three days has been a few points out of 3,000 or so — a change of perhaps 0.1 percent that could be reversed in a minute. This kind of trend is simply not the same kind of thing as the population shift from colder to warmer climates. In this section, I propose to examine enough examples, with graphs and a bit of real math, to help you identify *real* trends on your own.

A real trend

Suppose you have just been appointed manager of the utilities district for the small town of Tuckerville, an idyllic hamlet located in far northern California. Unlike most California towns, Tuckerville has, for obscure historical reasons, not only its own city-operated water service but also its own city-operated electric service.

Your job (for this example, anyway) is to forecast electric power demand for the next ten years. To help you make this forecast, you have all sorts of data from the last 40 years, including the number of households that existed in every year and the amount of electricity that each one used. You also have, from the town building department, a list of all the new structures that the town will allow in the next ten years.

Looking for a pattern

What do you do? The first thing might be to take the historical data year by year, plot it, and see whether you can find some sort of trend. Figure 8-1 shows you an example of this year-by-year plotting strategy. Actually, you can probably see two trends when you do so. First, twice the number of families means roughly twice the amount of electric power demand. On top of this is a slight, recent trend for power demand to drop a bit, as the most energy-using appliances, such as refrigerators, are modified in mandatory energy-conservation programs.

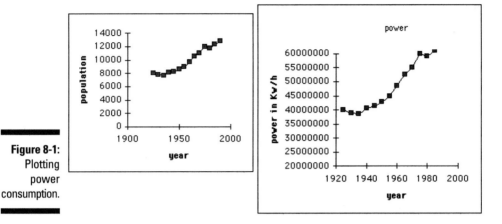

Figure 8-1:
Plotting
power
consumption.

You can analyze this information in two ways. In one way, you try to find a simple rule, perhaps of the form

Electricity demand = number of households × factor

One way to do so would be to pick a year, look at the number of households and the demand for electricity in that year, and find the factor by dividing. You could then look at a small sample, maybe four or five other years, to see whether the rule predicts electricity demand, at least within ten percent or so. Then you could make a little forecast with a reasonable level of confidence, as shown in Figure 8-2.

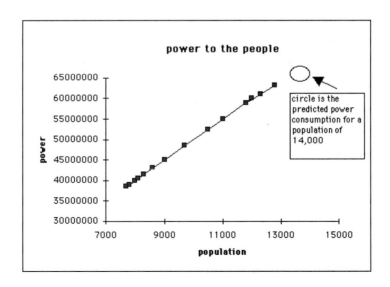

Figure 8-2:
The
Tuckerville
prediction.

Regression, anyone?

The proper mathematical way — at least the way many people believe to be the proper mathematical way — to find the trend would be to apply *regression analysis* on the household-versus-electricity data. This statistical technique draws a "best fit" straight line through some plot of the electric power use versus the number of households, as shown in Figure 8-3.

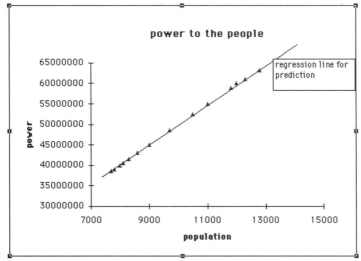

Figure 8-3:
Regression
analysis.

Regression analysis is built into scientific and business calculators, and of course is also available in every common computer spreadsheet program. Therefore, the actual computation of regression, which can be pretty tedious, is now handled automatically. The computation, however, is not the problem.

Regression and reality

The problem is in finding real economic and business situations to which regression analysis is the right technique to apply. Over the last ten years, I have been a consultant for several dozen biotechnology and computer-related companies in California and elsewhere. From the experience of hundreds of meetings in modern, well-equipped conference rooms, I regret to report that the business schools of America apparently turn out thousands of graduates every spring who think that you should apply regression analysis to data points in *time*.

Here's what I mean. In the case of Tuckerville, you can look at historical data and find some relation between the number of households and the demand for electricity. After all, any given household probably has a refrigerator, a certain number of light bulbs, a TV, some small appliances, and perhaps some electric heaters. That means that, at a minimum, more houses implies more electricity use, and a large sample of houses gives a household average to use.

Now you can also make a graph of electricity use per year. If the town were growing steadily over the years, it would look like a connection between years and power usage. But the real connection is between number of households and power usage, and this other plot just contains that information combined with the growth rate of the town.

But, in general, there is no *direct* connection between time and any other quantities except, by definition, your age. Time is not automatically connected to power usage, to unemployment, to sales of teddy bears or VCRs or tomatoes, to weather, to percentage of workers who are union members, or to height of the average citizen of Japan. All these quantities are linked to *something else* that is changing in time, not to time itself. To get to the bottom of things and find a valid prediction, you need to perform regression analysis on the original quantity (say *height*) and the something else that really affects it (in this case, *childhood protein intake*).

Not a real trend

To see how people get in trouble by failing to make the distinction between, for example, time versus sales and the real factors that influence sales, consider the case of Project X, a computer software company. Investors are currently suing the former managers of Project X, but I predict that the lawsuit will flounder. The managers weren't actively practicing fraud — they believed their own numbers as much as anyone. They were just *wrong,* wrong because they misapplied a bit of trend analysis.

Numbers don't lie?

Here's what happened. The programmers at Project X designed a semi-revolutionary business analysis tool. It took them two years, which was about nine months longer than was planned in the original budget sent to the investors. Oh well, these things happen. On a brighter note, when the public-relations people took early versions of Project X around to computer magazines, the press was very enthusiastic.

The first demonstrations resulted in lots of favorable press for several months. When the product was released, sales took off briskly, as shown in Figure 8-4. Ordinarily, businesses tend to report sales per quarter, but in a fast-paced racket like software, reporting every month is common.

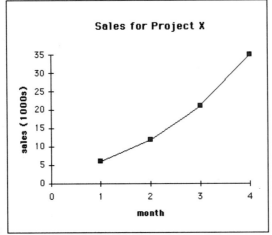

Figure 8-4:
Project X,
quick out
the gate.

And more data...

So far, so good. The problem developed after the product had been on the market for about four months. The first burst of pent-up demand relaxed, the first actual reviews (including serious product testing) were less than glowing, and the word from early users of the product was that it was a bit more difficult to use than expected. The reviewers and users tended to agree that the software was big and slow, and that, while it had plenty of cool features, it required lots of re-training to use effectively.

In months four through eight, the sales didn't grow quite at the rate predicted from months zero through four. Now the graph of sales by month looks like Figure 8-5.

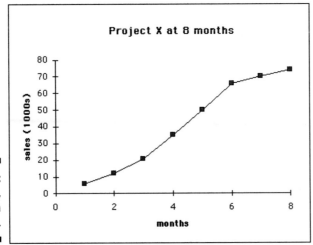

Figure 8-5:
Project X,
the saga
continues.

Prediction time

The managers of the company were obliged to call a meeting with the investors to tell them what future sales and profits were likely to be and to ask for more money.

At that point, the managers had three choices.

1. They could present a very cautious estimate of sales. If it's cautious enough, the estimate says that the project won't pay for itself for years. That means that it's time to fire up the old laser printer, spin the Rolodex, and start mailing out resumés all over the place since the investors are going to walk.

2. They could look at the data in Figure 8-5 and make a guess. In doing so, they would sound very unprofessional, not worthy at all of their MBAs, Countess Mara ties, little Italian shoes, and five-figure management salaries. That means that they're *not* going to guess, at least not consciously.

3. They could invoke regression analysis in a computer spreadsheet of sales data. At first glance, this looks like a hands-off, no-cheating, official-scientific-mathematical way to do things. Figure 8-6 shows what they predicted.

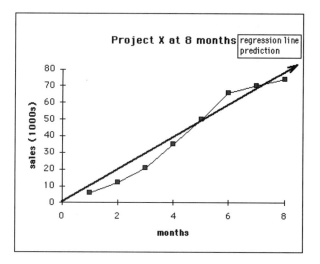

Figure 8-6:
Regression
applied to
Project X
sales.

And so the management of Project X confidently predicted that sales would double within a year.

Consequences

"Sweet dreams are made of this. . . . Who am I to disagree?" sang the Eurythmics. Well, that's what I'm here for — to disagree. To provide you with another bit of background information. Figure 8-7 shows the actual sales history, right up to the point where the whole thing blew sky high.

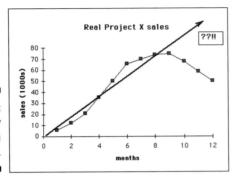

Figure 8-7:
Reality
intrudes on
mathematics.

What went wrong? Lots of things, and I'll enumerate them here:

- ✔ Setting up a regression study of sales versus time does not constitute making a model. Lots of factors influence software sales, and it would be possible to determine them, just as it would be possible to determine the influence of number of households versus power usage.

- ✔ Even with a model that takes lots of factors into account, a straight line is a pretty poor mathematical construct for summarizing sales results. Practically no examples in history show sales following a straight line for a very long time period. Never happened to IBM, didn't happen to General Motors, isn't true of Faroush's Corner Market. Wasn't the case here, either.

- ✔ Most fatally, this particular straight-line regression model doesn't take into account the most recent piece of information: that the sales increase is clearly slowing down in the most recent month. The way things work, this kind of slowdown is how sales tell you that they are going to level off and then turn downward.

In fact, sales went flat, then started to drift down, and then they dried up altogether. People lost lots of money. The investors, who felt that they had been duped by management at the big meeting, were in a particularly bad mood, hence the lawsuit.

Understanding Real-World Sales

Despite the mathematical convenience of regression and its miraculous capability to project small beginnings into great events later in time, sales of new products tend to follow a limited choice of curves in time, none of which looks like a straight line. Figures 8-8, 8-9, and 8-10 show a set of three representative curves.

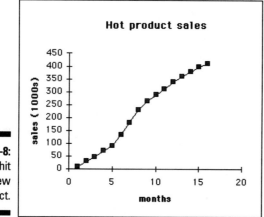

Figure 8-8:
A smash-hit
new
product.

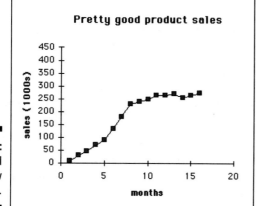

Figure 8-9:
A successful
new
product.

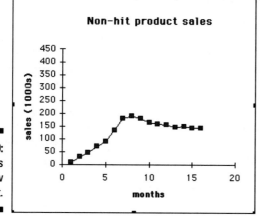

Figure 8-10:
A near-miss
new
product.

How math reality works

If you think about selling products into a market, you are at once confronted with two separate numbers: the size of the market and long-run market share. The basic logic is this:

- Every market is *finite*. Even if you had a product that gathered all the money in the world, that amount of money is finite. So at some point, even fairly far along, sales of *anything* level off. At some point, everyone who wants one has a pet rock, and pet rocks don't wear out.

- *Long-run market share,* the percentage of a market that your product commands, is frequently going to be less than you expect. In new technologies, you often have competitors you know nothing about — they appear as if by magic in the week that you introduce your product. Also, if you're very successful, you find that other companies can do a surprisingly good job of competing with you on the basis of price, features, distribution, or advertising.

Unless your product is a real dog, it will produce increasing sales as word of mouth, advertising, and press coverage prepare your market. And later, after everyone who needs a product like yours has had a chance to buy it, sales will slow down. Computer software companies, for example, have to have a version 4 ready to follow version 3 because sooner or later version 3 will saturate its natural market. In the case of Project X, the natural market just proved to be too small to justify a second version.

Thus the real shape of a sales curve will rise from zero, finally reach a peak level, and then settle down to a final equilibrium or long-run level. The only question facing a business is whether the long-run level, not the early peak, is good enough to justify supporting the product.

How human reality works

Sometimes people just reach right into their own sales curves, projections and all, and apply their own ingenuity to screw things up. Chrysler Corporation — a firm that has produced in the last decades enough business examples of decisions, both good and bad, to constitute an advanced course in management by itself — has provided a fascinating example of this phenomenon recently.

This example concerns the Plymouth Neon, Chrysler's answer not just to GM's Saturn but to a whole range of Japanese competitors. The Neon is not only a great small car, but it was designed in record time and brought in at a great price. Various economic pressures on the American middle class have induced a great appetite for cars priced under $10,000, and the first round of Neons were just flying out of the showrooms in early 1995.

To the financial staff back at Chrysler, this situation seemed like a crying shame. All these cars were blowing out of the dealerships, but the vast majority of them were basic Neons, without the deluxe package (nearly $3,000 more) containing a lovely assortment of high mark-up items with a profit margin almost equal to that of the basic car itself. So they ordered the production mix switched to "loaded" Neons and cranked out 200,000 of them to send to the dealers (*Business Week*, February 6, 1995).

Chrysler is a big company and has the resources to do sophisticated modeling, not just this straight-line stuff. Its models can predict changes in sales based on a $1/4$ percent change in the prime rate, in winter weather on the East Coast, and even on changes in competitors' prices. What they didn't predict was that *customers didn't perceive the basic Neon and the loaded Neon as the same product.*

There was a roaring appetite out there for less-expensive cars like the Neon. There was *not* an equivalent appetite for the same vehicle with overpriced pinstripes and half-measure amenities. Chrysler, having done a good job of forecasting demand for its product, turned the product into something different, at least in the eyes of its customers. At that point, all bets were off in the forecast. After an expensive lesson, Chrysler ended up discounting the fancy Neons and reorienting production toward the basic models.

Smoothing Data in Time

What you would like, in almost all attempts at forecasting, is a method that doesn't really make any assumptions about the trend that may be occurring. For example, the assumption that you're looking at a straight-line trend may often be wrong. One technique can help you see a bit more information about time-based data without introducing any bias in the results is *smoothing* the data. In its simplest form, this amounts to taking the data in bigger time chunks.

Time is on your side

You can see, for example, that if you were operating a national fried chicken restaurant chain, the individual day-to-day data from a single franchise would not give you the right information to see how you were doing on a national basis. You might want the numbers for sales in particular regions of the country, probably at time intervals of a month, but more likely at time intervals of a quarter of a year. For the most part, data on short time scales is just too jumpy to be useful in spotting trends.

Look at the Tuckerville data in Figure 8-11 and the Project X data in Figure 8-12. In these figures, a computer program with a time-smoothing function (it averages numbers over several months) is used to extend the plot by about a month. In both cases the prediction is pretty realistic, and this realism comes about because the time scale is appropriate. In the Project X case, for example, trying to draw a straight line through the sales data amounts to taking all the sales figures in time as equally informative. In fact, the last two months are just screaming for attention — they are signaling a dangerous potential downturn. They indicate that a change in the trend may be going on *right now*.

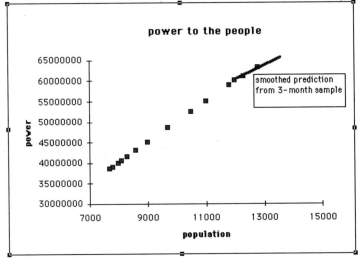

Figure 8-11:
Tuckerville
smoothing.

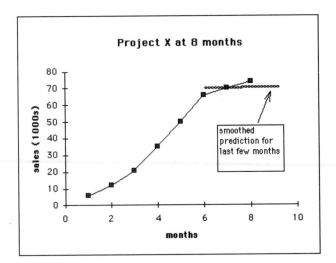

Figure 8-12:
Project X
smoothing.

Time at small intervals

One conclusion that can emerge either from an in-depth study of data or from about ten minutes' worth of thinking about individual cases is that there is always an *appropriate* time scale for looking at events. Consider these examples:

- ✔ If you had the temperature outdoors reported to you every 15 seconds, you wouldn't really be able to make a much better statement about global warming.

- ✔ Unless you can trade in and out of the stock market every 15 minutes, the standard reporting of financial news won't really do you much good. Actually, most studies of successful small investors show that they invest their money in the stocks of strong businesses, based on fundamental factors, and leave it there for years. There aren't many small-time "day-traders" who quit as winners.

- ✔ The government statistical tradition of reporting economic data month by month is almost comically counterproductive. Every December we hear that employment is up over the previous month, with extensive commentary pointing out that this is because of seasonal employment upswings from temporary Christmas-related jobs. Sheesh! I suppose it's part of the news being a faithful record of random-number events (temperatures, sports scores, traffic jams, lottery numbers, and so on), but it induces all sorts of concern where none is necessary. The appropriate time scale for employment and inflation news is probably six months, if not a year, rather than one month.

 Because you are usually a consumer, rather than a producer, of predictions and trend analysis, you owe it to yourself to think a bit about the right time scale for data reporting for each category of day-to-day information. This is especially true because most information providers are convinced that you want data every ten minutes even when a decade is the more relevant time interval.

And the other factor to consider is: What numbers go into trend models? Back in the 1960s, U.S. television manufacturers projected their future sales based on growth of the population and increases in the number of TVs per household. They got the number of TVs almost right, but neglected the interesting development that the TVs were going to be manufactured in Japan and Taiwan rather than in the U.S.

Similarly, the standard "leading indicators" about the economy that appear in the news, from consumption of paper pulp to durable goods, have become increasingly irrelevant in a world where the global economy, rather than domestic factors, is the source of most long-term trends. Perhaps this is an odd observation in a book that's mostly about aspects of mathematics, but for a big range of topics in trend analysis, common sense will serve you better than proficiency in calculation.

Chapter 9

Math and the Government

● ●

In This Chapter

▶ Understanding what an index is

▶ Thinking about inflation

▶ Thinking about unemployment

▶ Making your own economic index

● ●

*T*he United States government keeps track of all sorts of statistics related to the economy. These numbers are the source of nearly endless debate because they constitute the basis for policy decisions.

For example, if the rate of unemployment seems threateningly high, government agencies often try to do something to bring it down. Sometimes this takes the form of new government-sponsored job programs. Sometimes it takes the form of tinkering with interest rates (to make it easier for businesses to expand) through the Federal Reserve Board. And sometimes it just means that officials put pressure on the statistical groups responsible for unemployment figures to fiddle with the numbers until they make the apparent rate go down. So that you can understand what's happening in the newspaper headlines, this chapter will give you a bit of background on economic indexes and how they are computed.

Looking in the Index

An *index* is just a single number derived from a bunch of other numbers. Here's an example that I'll call the Breakfast Index.

The basics of indexing

The Breakfast Index has a set of elements that obviously relate to breakfast — not just some little Continental breakfast with two varieties of stale-looking croissants, either, but the real Iowa thing. Here are the elements:

- A dozen eggs
- A pound of bacon
- A one-day newspaper subscription
- A half-gallon of milk
- A stack of coffee-shop pancakes
- A can of frozen orange juice

When I assign realistic prices to these items and add up the numbers, I get an index, as shown in Table 9-1:

Table 9-1	The Breakfast Index in 1995
Price	*Index Item*
$1.29	A dozen eggs
$2.49	A pound of bacon
$0.75	A one-day newspaper subscription
$0.89	A half-gallon of milk
$3.19	A stack of coffee-shop pancakes
$1.17	A can of frozen orange juice

These numbers add up to $9.78, partly because some of the quantities are a little odd. For example, I hope you don't eat a dozen eggs and a pound of bacon every morning — a lot of concern about cholesterol levels is exaggerated or misplaced, but a pound of bacon a day won't keep the doctor away.

Now, suppose these numbers change over a few years. If these numbers are approximately accurate for 1995, I can either guess numbers for the year 2000 or I can get into my time machine and visit a supermarket of the future (in the interests of accuracy, that's what I'll actually do). The new numbers are shown in Table 9-2:

Table 9-2	The Breakfast Index in 2000
Price	*Index Item*
$1.36	A dozen eggs
$2.59	A pound of bacon
$1.25	A one-day newspaper subscription
$0.91	A half-gallon of milk
$4.59	A stack of coffee-shop pancakes
$1.37	A can of frozen orange juice

The new total is $12.07.

In an indexing scheme, you set the original total equal to the number one by dividing the total by itself. Then, for other years, you get an indexed number by dividing the new total by the original total. In other words

Index for 1995 = 9.87/9.87 = 1.00

Index for 2000 = 12.07/9.87 = 1.22

You can make a graph of this number over the years and use the graph to track an approximate price of breakfast. One of the conveniences of an index is that you can rattle off another number, the percentage change, just by looking at it. In this case, the index grows by 22 percent between 1995 and the year 2000. The big question for this and other index numbers is: Can you trust it?

The devil's in the details: index components

Now, just to see how indexes work in real life, look at the numbers one at a time.

 ✔ The price of milk is going nowhere, really. It's up barely 2 percent against a 22 percent increase in the overall index. That's because a complex set of agreements between the state and dairy farmers' organization regulates the price of milk in most states that have dairy cattle. The prices of many such agricultural products are fixed by strange, non-market price support mechanisms. *Non-market price support* means that the price is *not* set by supply and demand in the market. Milk prices are a political football in dairy states.

✔ The price of bacon and eggs also doesn't change much, because enough American adults have been bombarded with health-scare literature to think of these substances as practically toxic. For example, it's all the poor old State Egg Boards can do to keep sales level — raising prices very much is almost out of the question.

✔ Newspaper prices, in contrast to the prices of these humble farm products, went from $0.75 to $1.25. That's an increase of 67 percent! Part of the cost stems from increases in paper prices, and another part of it is the result of organizations trying to keep up revenues as newspaper circulation gradually shrinks (newspaper readership among the members of Generation X is half that of Baby Boomers).

✔ The coffee-shop pancakes also went up by a bunch. If this number were the only number included in the index, we would have a new Pancake Index instead of a Breakfast index and the index value for the year 2000 would be

Pancake Index for 2000 = 4.59/3.19 = 1.44

The Pancake Index shows a 44 percent increase, compared to an increase in the Breakfast Index of a mere 22 percent.

Suppose that I, as the statistician in charge of this operation, decide that the pancakes are distorting the overall numbers. After all, they're the only item not consumed at home, and they account for a big price increase. I thus propose the Revised Breakfast Index, which includes only eggs, bacon, newspaper, milk, and orange juice.

Now the new original total is $6.59, and the total for the year 2000 is $7.48. That gives you the following new index value:

Index for 2000 = 7.48/6.59 = 1.14

The lesson here is that by changing a single item in the index, I cut the increase practically in half, from 22 percent to 14 percent. The same thing is often true of other indexes as well. And this makes it possible to change the results significantly with a minimal amount of tinkering.

Inflation: What's the Real Number?

So it's fairly easy to change the value of an index if you can change one of the bigger components. Do you think it's likely that a given administration in Washington could resist the temptation to change the numbers if inflation seemed to be out of control? This, by the way, is a nonpartisan question — the record of both the Democrats and the Republicans on the issue of statistical integrity is about the same.

The consumer price index

The *inflation index* (also known as the *consumer price index* or *CPI*) is made up of a longer list of goods and services than the Breakfast Index that I discussed in the previous section, but basically it's the same idea. In principle, you just look up a whole set of prices in dollars in some particular year and then compare them to the prices for the same things in another year.

The catch is that the CPI determines a large assortment of different types of benefit and wage payments. For example, Social Security payments, different retirement arrangements, most union employment contracts, and businesses' internal payroll projections are linked to the CPI. If the CPI goes up by 3 percent, Social Security payments go up by 3 percent. In effect, the government is promising that retired people living on Social Security won't be ruined by a high inflation rate.

Fiddling with content

But what goes into the consumer price index? Should it include mostly raw materials, such as steel, petroleum, and wheat, or should it be finished goods, such as CD players, cable-knit sweaters, and shoelaces? In practice, it's something of a mix of both, but economists criticize the mix as being about 20 years out of date.

During the late 1970s and early 1980s, inflation in the United States was going at a furious clip by America's pretty restrained standards in this matter. As a big step-up in oil prices worked its way through the economy and raised prices of everything else, the *real* inflation rate (as opposed to the rate specified by the CPI) probably approached 15 percent per year, causing bank interest rates for home loans and so forth to hover near 20 percent.

This situation clearly posed a major political problem. The government would have to make drastic increases in payments on Social Security and other programs at a time of record budget deficits. If the CPI said that inflation was roaring along at somewhere between 12 and 18 percent (depending on the year), there would be a huge increase in payments.

The solution was quite simple: Start throwing out the components of the CPI responsible for most of the increase! It turned out to be politically expedient to adjust the official number for the inflation rate rather than face facts and report the true number.

At some point, for example, most housing costs disappeared from the index. It may come as a bit of a shock to readers in California and the larger cities of the East Coast, but as rents and mortgage payments started to approach 50 percent of take-home pay, these payments quietly dropped out of the index. That's right: The single element taking most of your income, and the element growing at the fastest rate, was removed from the index, producing an artificially low inflation rate.

It was the same as dropping pancakes out of the Breakfast Index; pick a key item and throw it out. Doing so makes it very hard to make accurate comparisons across several decades, but it solves lots of political problems in the short run.

Partly as a result of such statistical adventures, the real income of the American middle class has been dropping fairly steadily since 1973. When union wage contracts are renegotiated, wage increases in the contract are often determined by a formula that uses the CPI. But the CPI has itself been "fudged." Thus real increases in productivity are not being matched by real increases in income. Figure 9-1 shows a little graph that explains what's been happening in case you simply haven't noticed it in your own life.

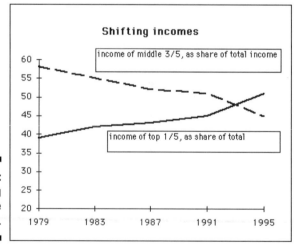

Figure 9-1:
Shifting income patterns.

Making time comparisons

Another inflation question has to do with comparing goods across time. The first Ford Pinto was designed to cost $2,000. Today, the lowest-cost car you can buy is probably the Geo Metro at about $8,000. But there's no comparing these two vehicles in terms of reliability or sophistication (I've owned both of them). A company couldn't sell a car as primitive as a Pinto in today's market, so we have no way of deciding what car is comparable to a Pinto decades later, the way a dozen 1980 eggs are comparable to a dozen eggs in the year 2000. The point that economists make is that the consumer price index tends to *overstate* price increases by ignoring improvements in quality. When quality is taken into account, many things are getting cheaper.

Computers are another, even more extreme example. About half the homes in America now have some sort of computer. But computers themselves have changed so much that comparing them across even four years is almost impossible. The current models are not just less expensive; they are phenomenally faster than the older models. For example, I'm typing this book on a Macintosh PowerBook that cost me a thousand dollars. In 1984, this computer not only wouldn't have been available at any price, but it would have looked like a prop from a *Star Trek* movie. Should computers be included in the CPI? If so, what would be the appropriate price to use: the price of the same machine or the price of a standard middle-of-the-road computer? At present, no one has a good answer to these questions.

Figuring the mix in the CPI now

Most economists now agree that the current mix of items in the CPI, despite all sorts of recent government work at fudging it downward in the last decades, actually *overstates* the inflation rate. In early 1995, the official rate is about 3 percent, and it's widely believed that it could get as low as 1.5 to 2 percent (see Figure 9-2). But after the experience of the last decades, changing the composition of the index has become harder.

The index as it stands now has more "old style" goods, such as large appliances, and fewer newer products (software, fax machines, and VCRs) that are getting less expensive every year. It also fails to account for real-world purchasing patterns — the new availability of generic rather than brand-name drugs is just one example. You could make up your own index based on your own spending patterns and come up with a projection of the amount of money that you will need to maintain the same lifestyle over the coming years, but since you're embedded in a world that relies on an official figure, your own superior information won't do you much good.

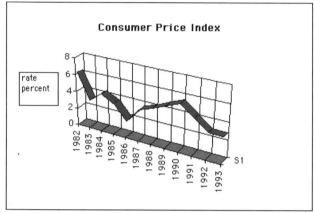

Figure 9-2:
"Official"
inflation
numbers.

Month-by-month CPI

One practice that may leave you puzzled is the monthly reporting of a CPI figure, followed by an annualized rate. For example, in the month of March, the CPI might change by 0.31 percent. What you will hear reported on the business news is this:

"Last month, the Consumer Price Index rose at a 3.7 percent annual rate, causing some alarm among government economists."

This practice of "annualizing" monthly figures by multiplying the monthly rate by 12 (for the 12 months in a year) makes every random microscopic blip in the statistics seem like a major crisis. A more appropriate time scale for CPI reporting would probably be every six months. Oh well, it gives 'em something else to use to fill up the time on TV news.

Unemployment Numbers: Should You Believe Them?

One of the things you may notice in this chapter is that when you formulate an index, you give up some detailed information in return for having the convenience of a single number to characterize a situation. You can describe inflation as a simple rate, but you lose information about which stuff is getting cheaper and which stuff is actually getting more expensive.

Similarly, describing unemployment by a single rate masks a variety of trends. The unemployment rate is statistically the percentage of the workforce collecting unemployment payments. Some little adjustments are made to this number, but basically that's it. Several factors make this figure quite misleading. Consider these examples:

- ✔ You lost your job in a depressed area, but because you can't sell your house, you keep on looking for work in the same town. After six months, you're still looking. At that point, you are considered not to be part of the workforce anymore — statistically, you've disappeared and so are no longer unemployed.

- ✔ Your electrical engineering job disappears as your company merges with another company that already has a big engineering department. The company offers to hire you 20 hours a week as a consultant at 60 percent of your previous hourly rate with no benefits. As far as the unemployment figures are concerned, everything's just ducky.

- ✔ The Federal Reserve pushes the prime rate up, thus raising mortgage rates to the point where the turnover of houses slows to a crawl. You quit working at the real estate agency and get a part-time job as a bartender. Congratulations! According to current employment accounting practice, you're in great shape.

Making up your own index

Rather than rely on government unemployment figures, you should probably make up your own index. Simply measure the number of full pages of employment want-ads in the Sunday edition of your local newspaper. Although not all job openings are reported this way, the want ads tend to comprise a fairly constant fraction of all jobs available. Because placement of ads follows trends around holidays (nothing is ever in the papers going into Christmas, for example), you should probably keep track of numbers of pages for a whole month.

Now take that number (pages per month) as the starting point of your index. In this case, you can start with the official government unemployment number and keep track of pages. Suppose, for example, that the *San Jose Mercury News* had 120 pages of want ads for employment in March 1995, and the government said that unemployment was about 7.5 percent.

If, by July 1995, the number of pages increased to 135, you could project a new rate as

starting rate × (original number of pages)/(new number of pages)

= 7.5% × (120/135) = 6.7%

In other words, when you see more want-ads in the paper, you know that the unemployment rate has gone down. With this index, you just take a starting point from the government figures and then modify it month by month using information that you collect yourself. Among other things, this will help you see more clearly what's happening in *your* area.

Amazingly enough, this simple procedure gives you about the same numbers as a much more complicated formula used by economists in different government statistical offices. This procedure, however, does a better job of tracking conditions in your own city. You can even modify it to produce a "good jobs" index, in which you look through the ads and count off all the jobs that you would consider good. This personalized statistic would probably have a lot more relevance to you than the single number that you're likely to hear in the news.

Lumping Numbers Together: The GDP

Again, the desire to characterize aspects of the economy with a single number can produce misleading information across the board. Inflation and unemployment are by no means the only examples. The *Gross Domestic Product* (GDP), the main index of economic progress, is a rich source of puzzles like the following:

- ✔ Does a $20,000 Toyota Avalon assembled in Kentucky make the same contribution to the GDP as a $20,000 Buick Regal assembled in Michigan?

- ✔ Does a Motorola cellular phone assembled from U.S. components at a Motorola plant in Singapore for sale in the U.S. make the same contribution as a cellular phone from a Motorola plant in Arizona?

- ✔ Does a U.S. software team consulting over the Internet and being paid in English pounds in an Isle of Man bank account contribute to the GDP?

Years ago, 96 percent of the U.S. economy concerned domestic production and consumption. Now it's seldom clear what's what. How does a Mercury Tracer (Japanese components, assembled in Mexico, sold at U.S. Mercury dealers) fit in? The whole issue is unsettled, even among economists. They agree, however, that trying to measure overall economic activity by the steel output and electric power consumption measure agreed upon in the 1950s is near-lunacy.

The point of this chapter, which could easily be expanded to book length, has been to convince you that

- ✔ Indexes are quite simple things.

- ✔ You can understand them.

- ✔ You can even make up your own indexes.

- ✔ Your own indexes may be more useful to you than the indexes you hear in the news.

"WHAT EXACTLY ARE WE SAYING HERE?"

Chapter 10
Insurance Math

· ·

In This Chapter

▶ Thinking about insurance

▶ Surviving life insurance hassles

▶ Dealing with risk insurance

▶ Math and deductibles

· ·

*I*nsurance is a somewhat confusing subject. One problem is that insurance companies themselves create a certain amount of the confusion as a marketing device. But the fundamental business principles of insurance are simpler than the rules for running a candy store.

Think of it this way: A true insurance policy is like a lottery ticket. If a low-probability event actually occurs, you get a certain amount of money. This would correspond to winning the lottery, except that in this case you have to die, get sick, or crash your car to collect. The key element here is that you buy insurance because the event (a *low-probability* event) would cost more money than you are likely to have. You don't have to insure yourself against events that don't cost much.

Your Money or Your Life!

Life! What a topic! I hope you're not finding this out for the first time here, but you're not going to live forever, at least not in your current form at the same mailing address. On the other hand, you have excellent chances of making it through the next few chapters at least.

What are your chances of making it through the year? They're not bad at all. Please inspect Table 10-1, calibrated for white American adults, who are in fact a prime insurance sales target for American companies. The rates for most minority groups, I'm sorry to report, are worse than these, in some cases 10 to 20 percent worse across the board. They're also harder to find in print.

Table 10-1	Death Rates for White American Adults	
Age	**Deaths Per Thousand Per Year**	
	Men	**Women**
30 - 34	1.8	0.9
35 - 39	2.6	1.5
40 - 44	4.1	2.3
45 - 49	6.8	3.7
50 - 54	11.1	5.5
55 - 59	17.8	8.2

This table doesn't take into account risk factors on the job. That's okay — you would be amazed at how small the occupational adjustment is from dentists to police bomb-squad assistants.

Insurance rates

Now, how does this information relate to insurance? I'll take the simplest kind, *term life* insurance. Term life is a little bet between you and the insurance company. The insurance company thinks that you'll live out the term. You would probably prefer that also, but you want to leave some money to people in case of your demise. If you put down a $100 bet of this type, what would be a fair payoff? (This assumes that you're dealing with a perfectly efficient, non-profit insurance company.)

If you are a 40-year-old male, the table says that your chance of dying in the next year is 4.1 in a thousand. That means that for every 4.1 dollars of insurance, the perfect payoff would be $1,000. To convert this to a more convenient number, the payoff on $100 would be

$$payoff = \$100 \times (1{,}000 \div 4.1) = \$100 \times (244) = \$24{,}400$$

Because term life insurance is usually quoted as a given rate in dollars per year for a given amount of coverage, I should convert this number to a rate per $100,000. I multiply the $100 by a conversion factor that gives me a $100,000 payoff:

$$premium\ rate = (100{,}000 \div 24{,}400) \times \$100 = \$410$$

That's a suspicious little number, isn't it? It suggests that you can read off the price of a $100,000 term life policy just by reading off the numbers in the table — it's that amount in hundreds. That's the case, actually.

Rate changes and policies

You may find that you can get lower rates than those quoted in the preceding example. If you are a nonsmoker and pass all sorts of predictive medical tests (normal blood pressure, standard cholesterol level, and so on), you may be at a much lower risk of dying in a given year.

You may also find many companies that offer much *higher* rates. Among other things, the companies have to cover their expenses. And beyond this, most people don't bother to comparison-shop for insurance rates very vigorously. Life insurance companies make lots of money because of this reliable phenomenon. It may seem to you that there must be huge variations in the number of deaths per year, given the number of disasters you see on TV news. But although earthquakes and hurricanes can cause big year-to-year fluctuations in property damage, they have little impact on overall death rates — the death rate tables stay pretty much the same year after year. Life insurance is the insurance companies' equivalent of keno in casino gambling — you can't come up with a system to beat it, and the casino can't lose. (See Chapter 17 for more information about the perils of gambling.)

Also note in the table that as death rates get higher with increasing age, the insurance rates get much higher. At some point in extreme old age, you have a 50 percent chance of dying in the next year. That means that you should have to pay $50,000 to get a $100,000 policy for the next year under ideal circumstances.

So what's the story on widely advertised life insurance policies for people over 65, the policies that accept everyone ("You can't be turned down!") and have low monthly payments? It's simple. The usual story is that the policy offers no payoff at all for the first few years you're enrolled and a tiny payoff later. Companies can offer insurance to everyone at low monthly rates as long as they take the precaution of paying out much less money than they take in.

Other kinds of life insurance

Term life is the kind of insurance that financial advisers tell you to buy. There are, of course, other kinds that the insurance companies would like you to buy — whole life, universal life, and so on. These plans have a built-in savings and investment component in addition to a term life component.

I'm not discussing the other kinds here because (and this is one of the objections the financial-counseling pros have) determining your real return on complicated insurance is phenomenally difficult. If you buy term life and put some money in a certificate of deposit at a fixed interest rate, you know exactly where you stand. In the simplest universal life policy I could find, the explanation of the yield took three pages. This is a ...*For Dummies* book, and I'm keeping it simple!

Risky Business

Life insurance makes a wonderful statistical model based on tables (*actuarial tables,* they're called) that document the probability of isolated events that can be predicted and proved. Term life is the very model of "real" insurance — meaning coverage of pure risk with no investment component — because both you and the insurance company know the risks involved and have the payoff written into a contract. Other insurance isn't quite the same.

Insuring cars

Car insurance, for example, is not like life insurance. For one thing, it's much more susceptible to scams. People can do all sorts of funny things on an accident claim report, so it's hard for the company to know what to expect from individual cases. For life insurance, you have to be dead to collect. For car insurance, all you have to do is know an auto body shop that practices "creative" appraisals. That means that car insurance companies have to build provision for fraud into their rates. Tables indicate your probability of getting into a car accident, but they don't have the direct relationship to insurance rates that the life insurance tables have.

Insuring against earthquakes

Earthquake insurance, a hot topic in California in 1995 (everyone's rates doubled) is another statistically difficult problem. Unlike death, which reliably creeps up to a certain number of people per year and apparently escorts them down a long tunnel towards a bright light, earthquakes happen infrequently. In any given year, there's a very small chance that a big section of a California city will suffer millions in damage.

And yet decades pass without *any* damage, and it's impossible to predict the dollar amount of damage when an earthquake actually occurs. In this case, faced with extreme uncertainty, the insurer's tactic is simply to charge tons of money (earthquake policies are usually more expensive than fire insurance) and hope for the best. If the next earthquake happens out in the desert, the policies represent pure profit. If the next earthquake levels Los Angeles, all the insurers will be bankrupt. The formula

insurance cost = (probability of damage) × (probable dollar amount)

contains *two* terms on the right side that can't really be calculated.

Insuring against floods

Floods, on the other hand, represent a case for which both terms are known. In some areas, there's a 100 percent chance of a flood every few years. And because the buildings have been flooded so often, it's easy to estimate the dollar damages (hey, they have records from the last one).

The Federal agency that funds disaster relief has recently gotten a bit grumpy about guaranteeing flood insurance in some areas. In effect, the agency is asking the rest of the country to underwrite the expenses of high-risk areas. They probably can't ask to have the Gulf Coast relocated for hurricane protection, but they are tired of rebuilding houses in California on the banks of the Russian River. Some of these houses have seen fish in the living room three times in ten years, and the fish came in through the bathroom window. The issue is whether a program is still insurance if it's supposed to provide protection against a sure thing. Or is it just a kind of advance purchase savings plan?

Deductibles

This subject is often a source of controversy inside companies, where employees vote for one health plan over another. Many people feel that a lower deductible (the amount you have to pay before the insurance starts paying the bills) means that you "get something" from an insurance plan. Often, the decision concerns a choice of plans that include such extras as eyeglasses versus "no frills" plans. Other people point out that the plans with higher deductibles and fewer extras are cheaper.

"Extras" and certainty

One point to ask is whether an eyeglass plan is really a topic in insurance. If you wear glasses, it's almost certain that they will get lost or get damaged or that the prescription will have to be revised over the course of four or five years. So a glasses purchase is not a low-probability event. Besides, purchasing eyeglasses usually does not wipe out your savings or force you to take out a crippling bank loan.

Thus, an eyeglass option in a health plan isn't really insurance. The company that offers the option knows how much it's going to cost, and it passes along that cost together with its internal administrative overhead and need for profit. The option doesn't protect you from financial risk; it just represents a somewhat more expensive way to buy glasses through a payroll withholding plan.

Big expenses and certainty

The Medicare program for the elderly is another example, played with much larger numbers than the eyeglass case, of something that isn't really insurance. A large fraction of people between the ages of 65 and 100 are going to need expensive medical attention. It's not a 1 in 500 chance; it's more like 4 chances in 5. Another reason this program isn't really typical insurance is that the beneficiaries aren't the people paying for it. It's not uncommon for someone in his or her late seventies to spend a week in a hospital to incur Medicare charges that total four or five times that

person's lifetime contribution to the Social Security system. The costs are paid by people who are currently working and paying into the system.

That's why Medicare reform isn't a topic in insurance reform. Medicare reform can have two aspects: control of medical costs as the largest single purchaser of medical services, and changes in the services provided. The situation doesn't resemble the circumstances of private medical insurance at all.

A deductible example

Consider these real numbers taken directly from the brochure of a national private medical insurer (I won't tell you which one exactly, but you would recognize the name). For the standard plan in my area, the options are

   1) Payments of $222 a month with a $500 deductible

   2) Payments of $257 a month with a $250 deductible

   3) Payments of $297 a month with a $50 deductible

It's a real tribute to peoples' confusion about this stuff that option 2 even exists. Mathematically, the choices are not at all comparable. I'm just going to discuss numbers in the first and last plans.

   1) In reality, you could have no medical expenses this year, you could incur some intermediate amount, or you could go past the $500 mark. On the $500 deductible version, you face one of these three situations:

      a) With no expenses, you pay 12 months × $222 = $2,664.

      b) With $360 in expenses, you pay $360 + (12 × $222) = $3,024.

      c) With $780 in expenses, you pay $500 (the deductible) + (12 × $222) = $3,164.

3) The same results for the $50 deductible option are as follows:

   a) With no expenses, you pay $12 \times \$297 = \$3,564$.

   b) With $360 in expenses, you pay $\$50 + (12 \times \$297) = \$3,614$.

   c) With $780 in expenses, you pay $\$50 + (12 \times \$297) = \$3,614$.

Check this out! At any combination of things happening or not happening to you, the high-deductible plan is much less expensive than the low-deductible plan. If you can accumulate $1,000 in savings to put somewhere safe to cover medical expenses, you can get an even better deal.

When you're shopping for private health insurance, the insurance company puts all this information in a little table that makes this calculation fairly easy. But the same rules are built into most group health insurance plans, too. In many cases, you don't have a choice — everyone else already voted before you got to your company. If you do have an option, though, make sure that you run through a deductibles analysis just like this one.

Chapter 11
Advanced Topics in Interest

· ·

· ·

The voice on the radio commercial says, "Call us! We'll help you consolidate your debts and get a payment you can live with." Lots of homeowners in America have over-consumed and have ended up with a collection of small but high-interest consumer debts that are increasingly difficult to pay every month. A TV here, a video camera there, skis, a fancy camera . . . at some point, all the little monthly EZ payments of $30 or so begin to add up to a not-so-easy $600 or so per month. And if one of the paychecks in a two-paycheck household is suddenly interrupted, it's panic time.

In this chapter, I'll look at a few specific cases relating to the finance of purchasing houses and other high-priced items. The title of the chapter is perhaps misleading — a *real* advanced topic in interest would be how Orange County, California (and many large businesses around the world) has managed to lose huge piles of money betting on interest rate changes in the derivatives market. However, my mission here is to provide aid and comfort to people who see themselves as math impaired, not to explain how committees of supposedly very bright people can make $100 million disappear in an afternoon.

Basic Refinancing

If you have a fixed-rate mortgage (a mortgage where the interest stays the same over the whole loan period) and interest rates drop, you may want to refinance the same loan. *Refinance* simply means replacing your old home loan with a new one while you keep the same house. You might resist the temptation to take out a much bigger loan, too, but that's the subject of the next few sections.

Table 11-1 repeats the mortgage table from Chapter 4. I'm justified in repeating it here because it's your guide to getting the right answer on the biggest math question you usually face.

Table 11-1	Mortgages (Again)	
Interest Rate	*Monthly Payment for 30 Years*	*Monthly Payment for 15 Years*
6%	$600	$843
7%	$665	$898
8%	$733	$956
9%	$805	$1,014
10%	$878	$1,075
11%	$952	$1,137
12%	$1,029	$1,200

Now, you face two considerations in deciding whether to try to rewrite the paperwork on your house:

🗸 **How big is the change in rates?** Every time you redo the papers, a whole assortment of people, from the lender to the title company to the appraiser, gets to charge you some sort of fee. In the case of the lender, these are the notorious *points* added to the loan amount and other loan-origination fees (documentation fees, for example).

🗸 **When do you plan to move?** Your monthly payment may go down, but you may not make back enough to justify all the fees you paid if you move in two years.

I'll take numbers directly from the table for simplicity. You can scale them to your own loan amount by adjusting for an amount other than $100,000 (there's a long discussion of this process in Chapter 4). Here are typical steps in deciding whether a refinance is in order:

1. Compare the rate of your mortgage to the current fixed-mortgage rate. Just for something to consider, suppose that the rate is now 7 percent for a fixed mortgage, and you are holding paper at 9 percent.

2. From Table 11-1, a $100,000 loan amount would mean a drop in monthly payments from the $805 you are paying now to $665 a month at the new rate. Except that's not it exactly, because the lender will charge you fees for doing the refinance: I'll pick $3,100 as a typical number for a loan of this size.

3. The actual new monthly rate, instead of $665 per month, has to be adjusted because the loan amount will now be $103,100 (the original $100,000 plus the $3,100 in fees). The new rate is

$$\$665 \times (103{,}100 \div 100{,}000) = \$687$$

This still looks pretty good.

4. When you start the new loan, your first payments go almost exclusively to interest. If you sell the house after two years, your loan balance is $100,930 (finding the remaining balance after a number of payments calls for a financial calculator with a BAL — balance — key or equivalent). You saved

$$24 \text{ months} \times (\$805 - \$687) = 24 \times \$118 = \$2{,}832$$

by making lower payments for two years, but you actually owe more on the loan than you did in the first place. With a $100,000 loan amount at 9 percent interest, you would owe $98,569 on the loan after two years.

What do you make of all this muddle? There are really two conclusions. In this case, if you stay in the house for five years, the difference between the balances owed shrinks to about $1,200 (you are catching up faster at the lower interest rate, although you still owe a bigger balance on the new lower-interest loan), and you save $7,080 in monthly payments.

The other issue is that if the interest-rate drop is large enough, you can justify a shorter-term refinance. However, you are very unlikely to get a big drop unless you are the last person in the world to rewrite your ghastly 14 percent loan from the 1980s.

How do you decide what to do? This is a lot of calculation! So you should ask the mortgage broker to prepare a year-by-year table of the savings from the lower monthly payments and the total loan amount owed for both the refinanced and non-refinanced cases (don't worry about being a bother — loan brokers have computers for this). In other words, don't focus on the lower payment alone. This type of table shows you how long you have to stay in the residence to break even on the deal.

Dangerous Refinancing

There's another possible refinance option when the interest rates change by at least a few percent. Suppose the 9 percent to 7 percent drop happens again. This time you see that the rates have gone down and wonder how much more money you can borrow and still make the same monthly payments. For the cases in Table 11-1, it's simple and straightforward. You could increase the loan amount from $100,000 to

$$\text{new amount} = \$100{,}000 \times (805 \div 665) = \$121{,}052$$

while keeping the same monthly payments, just because the interest rate has gone down. (See Chapter 4 for the rule for figuring out loan amounts other than $100,000.)

So even with $3,100 in fees, you can clear $18,000 on the refinance and still be making the same monthly payments that you've been making all along. Is this a good way to buy a car? You could, from the looks of it, cruise home in a new Whatzis 900 without impacting your monthly budget at all.

The catch, of course, is that you don't impact your monthly budget but you do impact your net position in life. Four years later, your outstanding loan balance is $115,573. If you hadn't refinanced, your loan balance would be $96,858. That's a difference of $18,715 that, for all practical purposes, you still owe on your $18,000 car. Hmmm. If, of course, you live in an area where real estate prices are rising rapidly, you can get away with this dodge. If real estate prices are level or going down, you have put yourself in a hole big enough to bury yourself, the now-depreciated car, and a few big trash bags full of money.

The Credit Card Defense

I'll keep things simple by using the same numbers. You start with a $100,000 loan, but you have accumulated $18,000 in debt on credit cards. To keep things even simpler, I will assume that all the credit card debt is at 18 percent interest. The cases for comparison are 1) Don't refinance and make minimum monthly payments on the credit cards, or 2) Refinance and pay off the cards.

Here's how to determine which is a better deal — it's just a matter of comparing total payments and final results in these two cases.

Case one: Don't refinance

Every month, you make the minimum payment on the cards. That's about $1\frac{1}{2}$ percent (I'm going to make an approximation that's correct to within a few hundred dollars) of $18,000, or $270. You also make the payments of $805 on the $100,000 loan (fixed rate of 9 percent for 30 years). After two years, you have paid out

$$24 \text{ months} \times (\$805 + \$270) = \$25{,}800$$

Your debts are now

- ✔ still $18,000 on the plastic. At 1½ percent per month, you are just covering interest.
- ✔ $98,570 on the home loan.

Your total paydown of all debts over the two years is thus $1,430 for the $25,800 you have paid. Does that sound discouraging? It should.

Case two: Refinance, pay off credit cards

If you refinance and pay off the cards, you now have a $121,100 loan at 7 percent (I'm throwing in the $3,100 in loan fees for the sake of realism).

You have a balance of $118,550 on the home loan after two years. That's if you just make the $805 payments on the home loan. Your credit card debt has just turned itself into a home loan debt.

But you have an opportunity in this case. Consider making payments of

- ✔ $805 on the 7 percent home loan, the new loan payment on the larger home loan amount
- ✔ $270 per month extra as a payment against loan principal (almost every loan arrangement accepts these extra payments)

Making the extra home loan payments, you would then have a home loan balance of a little less than $110,000. And you do this just by making the same payments that you were making in the first (no-refinance) case — remember, you were paying $805 on the house and $270 on the cards. Keeping up the same payment schedule ($805 + $270) in the refinance case, you would have paid off nearly half your mountain of credit card debt in two years and would be more than clear of it in four years. But note this crucial condition:

To really get rid of credit card debt with a refinance, you should be making extra home loan payments in the amount that you were paying each month on the cards.

You transfer the debt to an interest rate that's easier to handle but also transfer it to a longer payback period. Maybe you can't make the whole credit-card-equivalent amount every month, but you should think of your old card balance as a debt that's still out there, waiting to be paid. It just isn't growing as fast.

How to go broke

In case the preceding section didn't demonstrate it clearly, here's how to utterly ruin yourself. First, run up a big pile of credit card debt. Second, refinance and make only the new home loan payments. Third, realize that you now have a new bunch of squeaky-clean credit cards and head on down to the mall to run them up to the limit again.

Now you are completely destroyed, unless someone gives you enough money to get level again. You owe a pile of money at a higher interest rate, you have used up your refinance opportunity, and you owe a bigger total than ever.

I never thought I would be writing anything like this in a math book. But when I told various people about this project, questions about fixing credit cards with a home loan came up almost immediately in the group of people I used as chapter testers. One championship case from real life involved a couple refinancing $35,000 of credit card debt with a loan that wiped out all equity in their home.

Equity, as your parents should have told you, is the difference between what the house is worth now and the amount of the loan. If you don't have any equity when you sell the house, you get no money back from the deal. The happy couple in my example then went out and ran up another $20,000 on the newly laundered cards. They'll probably be looking at formal bankruptcy in another year or so.

I have every confidence that this isn't even an extreme case. Credit counselors probably hear worse stories all the time. So here's some gratuitous advice:

- ✔ Get a copy of Eric Tyson's *Personal Finance For Dummies* and read it, especially the part about credit management.
- ✔ In the meantime, pay as much as you can on high-interest credit card debts and try not to make any major purchases for a while.

Debt and Taxes

Interest charges on home loans are tax deductible. Interest charges on credit cards used to be deductible, but then the tax code was revised. An executive second home at the seashore is deductible, and the Mercedes in the doctor's office parking lot is probably a deductible business expense, but the video camera you put on MasterCard is not seen in the same light.

Looking at the preceding example, you see that taxes don't actually change the bottom line much in these loan examples. You can shift the credit card balance to the home loan, but to get rid of the debt in a reasonable amount of time, you have to make payments against the principal of the loan. Payments against principal are not deductible.

The only way to gain a bit of a tax advantage is to shift to a shorter-term home loan. If you end up with a $121,100 loan at 7 percent at 15 years instead of 30, your payments will be as follows (refer to Table 11-1):

$$\$898 \times (121{,}100 \div 100{,}000) = \$1{,}087$$

That's very close to the $805 + $270 that you were paying in both the idiot's delight plan (keep the cards and keep the 9 percent loan) and the 30-year refinance (pay the 7 percent loan and pay $270 extra). But now almost the entire amount for the first few years is tax deductible (it's nearly all interest). After two years, your loan balance is $111,120, showing that you have made some headway into the total debt despite the large fraction of the payments that's just interest.

Part III
High School Confidential

In this part...

1 t should not come as a surprise to you that the U.S. makes a pretty poor showing in math education for an industrial country. What you may not know is that the U.S. has some company in the back of the class. Typically, our numbers are pretty close to those for Canada and the U.K. Actually, the math test numbers for some individual states (Minnesota and Iowa are prime examples) are significantly better than those for most of Europe. What, however, is the connection between lower math scores and the English language? Having spent years in the educational trenches teaching statistics, I have a theory about Where It All Went Wrong.

In England, it's fair to say that for many decades mere applied math was not something in which a real gentleman would take a full-time interest. Engineering education was left to the second tier of schools. As a result, the English mathematical system developed an orientation toward a sort of sterile intellectual competition. American and Canadian practice, I think in imitation of the way things were done at Oxford and Cambridge, developed an emphasis on trick exam questions as the heart of math education.

That approach sometimes calls forth great results from the 0.5 percent of the population that has superior math ability. It also produces millions of people every year who graduate from secondary school with no memories of math except "I went in there, looked at the final, and didn't recognize *anything*. I don't know how I passed — I just put down a bunch of guesses." The standard school approach to math is something like teaching English literature using the Sunday *New York Times* crossword puzzles as the main study aid.

Well, that's not where this work is headed. I don't have to consider whether I am preparing you properly for a professional mathematical career. I can tell you what the problems are and how to solve them. It's a simple cookbook approach, and it works. This is first aid to be used in the field, not a tour through medical school.

Teacher's note

I am the descendant of generations of secondary and university teachers, and I think that teachers are the finest people in the world. I wish that teaching paid well enough that I could afford to do it again. I know high school teachers who put in countless unpaid hours helping students and buy thousands of dollars' worth of classroom supplies every year. I am not questioning anyone's sincerity or ability. In this context, the measurable lack of results from all this effort on a national scale suggests that *we're just doing the wrong things*.

Here's a little cultural perspective. I have a friend from South Korea who works as a computer-chip designer. We were at a reception at a trade show in Boston, standing around with a large group of technical people, one of whom mentioned that her secret fantasy was to date a football player. My friend remarked, "In Korea, girls fantasize about dating electrical engineers. Which culture do you think has the right fantasy for the next hundred years?"

Chapter 12

What They Were Trying to Tell You in Algebra

*A*lgebra in the high school sense is simply a set of rules for solving problems involving numbers. If you had trouble with algebra, the trouble probably came from one of three sources:

✔ You had a hard time deciding which rule to apply to which problem.

✔ You had a hard time translating a problem from words into symbols.

✔ You had a hard time seeing the point of a bunch of apparently unconnected tricks.

Furthermore, algebra seems to be introduced to people at the exact point at which, for developmental, hormonal, or other reasons, their attention span for abstract symbolism has been reduced to a fraction of a second. Personally, I think that people should be taught algebra when they're 9 and social studies when they're 30. What's the point of learning about boring old Warren G. Harding in U.S. History if you're too young to hear *exactly* how he died in a hotel room in San Francisco? (*Hint:* He wasn't alone and apparently wasn't all that boring.)

I'm going to cover some highlights of the material that you may have missed while your social life was evolving in ninth grade. It's actually pretty interesting, and a mere handful of tricks solves most of the problems that you encounter in real life.

The Numbers

For reference, I'm just going to list the kinds of numbers that underlie ordinary algebra first. Once I start talking about *xs* and *ys*, one of these kinds of numbers will get "plugged in" for the *x* or *y* value.

Integers

Integers are whole numbers — 3 or 17 or 365. "God made the positive integers; everything else is the work of man" is a quote from the famous German mathematician Leopold Kronecker. That means that 1, 2, 3, 4, and so on were already here in some sense, corresponding to the number of antelopes in a field or coconuts in a tree, before we even started to think about them. Right there at the end of our arms are little sets of integers that we can actually wiggle. Hey, they're even called *digits*.

The first human-invented integer was zero. There is reason to believe that zero first appeared as a notation for an empty space on a counting board (see Figure 12-1). When zero is used this way, it's possible to make a calculation on paper have the same form as the more familiar (in those times) calculation on a counting board.

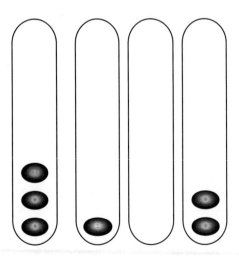

The number 3,102 on a counting board with beads. It could be represented as

Figure 12-1:
The origins
of zero.

III   I   ·   II

with the dot standing for "empty" or zero.

We know who invented the lightbulb (all six inventors, actually), but we don't know who invented zero. It's known that it happened twice, independently, once in Central America for the Mayan counting system and once in India (some people claim that it may also have been invented in China at nearly the same time).

Finally, there is the man-made invention of *negative* integers. These numbers can be viewed as either an unnecessary idea for subtraction or, better, an extension of the idea of numbers lying along a line.

Take the number (–4), for example. You can add it to 6 to get the following:

$$6 + (-4) = 2$$

and you get the result 2. Why? Because adding a negative number is the same as subtracting a positive number. And what, you may ask, does subtracting a negative do? Well, as this example shows, subtracting a negative number is the same as adding a positive number:

$$6 - (-4) = 10$$

When you look at the numbers all strung out along the line in Figure 12-2, you see the rationale for the adding and subtracting of negative numbers. In the arithmetic operation, you are moving back and forth along the line. If you start at 2 and add 3, you move on out to 5. If you start at 2 and add (–3), the minus sign tells you to move to the left. *Add* means "move to the right" and *subtract* means "move to the left," but a negative sign on the number reverses this direction.

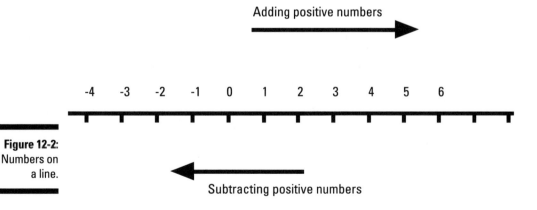

Figure 12-2:
Numbers on
a line.

Rational numbers

Rational numbers, which as a term include the whole numbers and all the fractions of whole numbers, are the next human invention — although referring to the previous example of antelopes on the plains, it should perhaps be noted that the concept of a half antelope was entirely familiar to lions millennia ago. In fact, the proverb "Half a 'lope is better than none" actually originated among lions.

Examples of rational numbers are

- $^1/_4$
- $^3/_8$
- $^{56}/_7$
- $^{59}/_7$
- $^{16}/_{64}$
- $^{355}/_{113}$

Just taking a look at these in order, only the first would have been considered a "decent" fraction in the ancient world. An Egyptian mathematical document called the Rhind Papyrus has a whole host of formulas for dealing with fractions, but all of them assume that the fractions will be expressed in a 1/*something* form.

For example, the next number would have routinely been written as

$$^3/_8 = {}^1/_4 + {}^1/_8$$

This fraction, in this exact peculiar usage, turns up later in history. When the Republic of Venice essentially hijacked the Fourth Crusade and directed it against Constantinople instead of Palestine, the Doge of Venice triumphantly proclaimed himself "Lord of a Quarter and Half a Quarter of the Roman Empire." That does actually sound more impressive than "Lord of $^3/_8$ of the Roman Empire" or the even less compelling "Lord of 37.5 Percent of the Roman Empire, Discount Taken at the Register."

The next rational number in the list, $^{56}/_7$, is not just a rational number; it's an integer, since

$$^{56}/_7 = 8$$

Just by taking this as a clue in a division problem, you can probably see that the next number in the list is a bit top-heavy, because

$$^{59}/_7 = 8 + {}^3/_7$$

Another fraction that "needs work" is $^{16}/_{64}$. By a lucky guess, you may be able to see that 16 divides evenly into 64. If you don't know this, you can still reduce the fraction to its basics by dividing the top (numerator) and bottom (denominator) by 2 to get

$$^{16}/_{64} = \frac{(8 \times 2)}{(32 \times 2)} = {}^8/_{32}$$

and then

$$^8/_{32} = \frac{(4 \times 2)}{(16 \times 2)} = {}^4/_{16}$$

and then

$$^4/_{16} = \frac{(2 \times 2)}{(8 \times 2)} = {}^2/_8$$

and then

$$^2/_8 = \frac{(1 \times 2)}{(4 \times 2)} = {}^1/_4$$

What a complete waste of time! Fractions are still around because you can do so many little computational tricks with them, and because the publishers of books about sewing, cooking, and home improvement keep insisting on traditional measurements like $3^1/_4$ yards of fabric, $2^1/_2$ cups of flour, and $1^1/_2$ by $3^3/_4$ foot pieces of plywood. Even the stock market lists share prices as \$42$^3/_8$, a distant echo of "pieces of eight" and the origin of the term "two bits" for a quarter.

It's better, if the word *better* means anything at all, just to convert all this stuff to decimal fractions. In decimal fractions, $^{16}/_{64}$ becomes 0.25, which you can probably recognize right away as $^1/_4$, and if you can't recognize it, you can divide 1 by 4 on a calculator to remind yourself.

The last rational number shown here is $^{355}/_{113}$, a very good approximation for π (pi). In real life, on computers and calculators and paper, you always have to settle for a *rational approximation* to special numbers like π, because neither you nor your computing device can deal with an infinite number of digits. How good is this particular approximation? If you use it to calculate the moon's orbit, you get the circumference of the orbit wrong by about the width of a fireplace.

Irrational numbers

These numbers can't be expressed exactly as fractions or as integers. One kind of irrational number can be expressed in the form

$$a^b$$

where both a and b are rational numbers. The famous example of this expression is the square root of 2. For the square root of 2, a is 2 and b is $^1/_2$, a perfectly respectable rational number. By the definition of these things in algebra notation

$$\sqrt{2} = 2^{1/2}$$

This number starts out 1.41421356237309505... and continues forever. It's an irrational number, meaning that you can't express it just as a fraction. This class of irrational numbers can be constructed by a limited number of algebraic operations, though, since it has to be expressible in the a^b format.

The truly weird kind of number is called *transcendental*. You can't put it into a^b format, and, in fact, you can't even express it in a finite number of operations. The transcendental number that everyone knows is π, but another number

$$e = 2.71828182845904524...$$

turns up in a huge variety of calculus formulas. Most of those formulas are concerned with growth, which is why this strange transcendental number is at the heart of computations in compound interest. There's something pretty strange about calling on this spooky little number to calculate credit card interest.

The Rules of Algebra

One of the nice things about algebra is that you already know a lot of it, in the sense that most of what you expect to be true of numbers is also true of numbers represented by symbols.

Using letters to stand for any of the classes of numbers listed in the preceding sections (they're the *real* numbers, as opposed to imaginary numbers described in Chapter 25), you have a small set of rules:

$$a + b = b + a$$

$$a + (b + c) = (a + b) + c$$

$$a \times b = b \times a$$

$$a \times (b \times c) = (a \times b) \times c$$

$$a \times (b + c) = a \times b + a \times c$$

You can look at these equations with numbers plugged in for letters to convince yourself that for the ordinary numbers you can display on a calculator, these rules are always correct.

$$4 + 5 = 5 + 4$$

$$1 + (2 + 3) = (1 + 2) + 3$$

$$3 \times 7 = 7 \times 3$$

$$3.14 \times (5.3 \times 2.7) = (3.14 \times 5.3) \times 2.7$$

$$52 \times (34 + 13) = 52 \times 34 + 52 \times 13$$

Other algebras

The reason that textbooks go on at great length about these rules for real numbers is that they aren't necessarily true for other potential types of algebra. Mathematicians can define an algebra with special objects A and B and an operator $\oplus$, where, for example

$A \oplus B$ is *not* equal to $B \oplus A$.

A lot of the most important mathematics of the 20th century concerns types of algebras in which the familiar rules that apply to real numbers just don't apply at all. It may sound strange, but these rather other-worldly algebras are important in many practical applications in physics. Elementary particles don't care *what* the story is on real numbers — they've got their own rules.

Manipulating number symbols

For the purposes of this chapter, the rules that I just listed mean things as self-evident as

$$4 \times 5 = 5 \times 4$$

which you probably haven't had much doubt about lately. You probably figure that $2 + 3 = 3 + 2$ also. Because the algebra in this chapter concerns only real numbers, anything on both sides of an equation is a real number. That means that you can do both of the following:

- ✔ Multiply both sides of an equation by anything you like.
- ✔ Add or subtract anything you like to both sides of an equation.

If you are confronted with an expression like this:

$$4x^2 + 3x = 12x$$

you can

1. Subtract $3x$ from both sides, giving $4x^2 = 9x$.
2. Divide by x on both sides, giving $4x = 9$.
3. Divide by 4 on both sides, giving $x = {}^9/_4 = 2{}^1/_4$.

(Actually, step two is a bit tricky. Another solution of the original equation is $x = 0$. It's always worth doing a fast check to see if $x = 0$ is a solution.)

What you're usually trying to do in algebra is to get to the definition of some unknown quantity in terms of a simple number.

By the way, a notation in which individual letters stand for the unknown quantities took a long time to evolve, and the now-familiar notation

x^2 for "x times x"

came later still. The x, y notation for unknown quantities was invented in the late 16th century in France by François Viete — one of the first people to write an equation that you or I would recognize as algebra. Before that, it was pretty much standard in European and Arabic math to write equations that were the equivalent of statements like

Profit = Revenues – Cost

keeping whole words as names for the quantities to be determined.

Problems, Problems

Most of the real-life applications of algebra consist of converting problems in words into problems in symbols, since it's usually easier to solve the symbol version of the problem. The emphasis on this aspect of algebra is simple to explain: You are very seldom confronted with quadratic equations (equations that have x^2 as the highest power of x in the equation), but a simple bit of algebra crops up almost every day.

Problem 1: Getting there

At 70 miles per hour, your car gets 25 miles per gallon. At 55 miles per hour, your car gets 30 miles per gallon. You are 270 miles away from Las Vegas on an empty stretch of desert road with no gas stations along the way. From long experience with your car, you can tell from the needle on the gas gauge that you have ten gallons left. It's two in the afternoon. When do you get to Las Vegas?

The first thing to do in this type of problem is separate it into cases. Take case 1 first, the 70 mile per hour case. The real question here is this: If you get 25 miles per gallon, how far can you go on ten gallons? From practical experience, you can say that on the first gallon you go 25 miles, on the next gallon you go another 25, and so forth until you've used up all ten gallons. In other words

miles you go = 25 (miles/gallon) × 10 gallons

= 250 miles

The gallons in this multiplication cancel (you have gallons on the bottom of one units-fraction times gallons in the other term), just as if they were xs or as or bs. In notation, you can say that the formula is just

$D = m \times g$

where

D = distance you go in miles

m = mileage in miles/gallon

g = available gallons

What you see from this demonstration is that if you start at two in the afternoon and drive at 70 miles per hour toward Las Vegas, you'll get there only after hitchhiking the last 20 miles. That could be a long time.

So apply the formula to the second case. If you drive more slowly, you get 30 miles per gallon as your mileage. (Actually, these numbers are fairly realistic.)

$D = m \times g$

300 miles = 30 miles/gallon × 10 gallons

You can make it all the way to Las Vegas if you drive more slowly. That realization gets you to part two of the question. At 55 miles per hour, how long will it take you to go 270 miles? The right formula is

distance = miles per hour × hours

$D = s \times t$

where

D = distance in miles

s = speed in miles per hour

t = time in hours

What you *want* is a formula that says that t = something because you're looking for the time. But this formula will do, since you're left with a way to get t if you put in the two available numbers.

Plugging in the numbers, you get

270 miles = 55 miles per hour × t

Multiply both side of this by *hours,* divide by *miles* to clear the units, and the equation reads

270 hours = 55 × t

If you divide by the number 55, you get

4.91 hours = t

It's going to take you almost 5 hours, as 4.91 is nearly 5. In other words, 4.91 hours is 4 hours plus

0.91 hours × 60 minutes/hour = 54 minutes

Looking at the mileage figures, you see that you risk running out of gas a few miles short of the bright lights of the fabulous Strip if you try to go any faster than this. No hurry, though — see Chapter 17 for a rather stonily practical chapter on gambling.

Problem 2: The ridiculous age problem

You saw problems like this in high school, and you saw problems like this on the SAT or ACT test if you took one of them. And then you never saw them again, except perhaps in puzzles. You probably could have a very successful career in business without being able to solve this kind of problem at all. But because this part of the book aims to explain what they were trying to tell you in high school, I feel obliged to recapitulate.

George is four years less than three times Steve's age. Steve is now 12. How old is George?

Nothing like the experience of looking at the clock, seeing that you have three minutes left to complete an algebra test, and being confronted with this sort of question. Ever meet someone who claimed to be your age times the square root of π plus two? Anyway, to put it into symbols, just say that

S = Steve's age

G = George's age

Three times Steve's age minus four years gives you George's age. That is, George would be exactly three times Steve's age, but the problem says that he's four years younger than that. In symbols, that's

$3S - 4 = G$

Taking the multiplication statement first, if you plug in the number 12 for S, you get

$3 \times 12 - 4 = 36 - 4 = 32 = G$

George is 32.

Actually, problems like this are just exercises in translating a set of conditions into algebraic statements for solution. Often the problems are somewhat contrived, but practicing symbol-to-word translations is more than half the battle in algebra. The rest of the battle is just practicing all the "subtract from both sides, multiply by both sides" stuff.

Problem 3: The almost-ridiculous mixture problem

Here's another classic problem, repeated down through the decades in textbook after textbook. Around 1980, some algebra textbooks started featuring lavish use of color printing, but no one seems to have thought of updating all the old problems.

A five-pound can of mixed nuts contains cashews and peanuts. It costs $30. If peanuts cost $4 a pound and cashews cost $9 a pound, what is the amount of each in the mixture?

In this case, it's almost simpler to translate everything in the problem into notation right away than to try to think about the problem in these terms.

First, you can say that the number of pounds of peanuts in the can is

number of pounds of peanuts = p

since you don't know what the number is and the letter p is pleasantly evocative of peanuts. Having done so, you have also established the number of pounds of cashews. No, it isn't c. It's

number of pounds of cashews = $5 - p$

because the problem states that it's a five-pound can. Everything in the can has to add up to five pounds, so

peanuts + cashews = 5

p + cashews = 5

cashews = $5 - p$

The next expression to translate is the total cost information.

Thirty dollars has to come from p pounds of peanuts at $4 per pound and $(5-p)$ pounds of cashews at $9 per pound. Gaily disregarding the units of pounds and dollars for the moment, you can write

$$30 = p \times 4 + (5-p) \times 9$$

or, a bit more cleanly, you multiply 5 by 9 and $-p$ by 9 to get:

$$30 = 4p + 45 - 9p$$

Resorting to mere algebra, you add $4p$ and $-9p$ to get

$$30 = -5p + 45$$

Subtracting 45 from both sides gives you

$$-15 = -5p$$

Multiplying both sides by -1 gives you

$$15 = 5p$$

Finally, you divide both sides by 5:

$$^{15}/_5 = p$$

which, reduced, is

$$3 = p$$

The can contains three pounds of peanuts, which in turn means that it contains two pounds of cashews.

As a quick common-sense check, look at the numbers involved. The average cost of $30 worth of stuff in a five-pound can is $6 a pound (it's just $30 divided by 5). The peanuts are $4 a pound, and the cashews are $9 a pound. Because 4 is closer to 6 than 9 is, you can reasonably expect to see more peanuts in the mix than cashews. So the answer makes sense, which is always nice. Sometimes these problems are pointlessly arbitrary, but you can usually use a little common sense to check the answers.

Logarithms

I'm going to shift gears for a moment and talk about another use for algebraic notation. Suppose I formulate another problem in x that requires a bit more effort than plain arithmetic. Here it is:

$$8 = 2^x$$

You can poke around, try some numbers, and convince yourself that x can equal 3 since

$$2^3 = 2 \times 2 \times 2 = 8$$

And 2^4 (just putting one more 2 in the stack) is 16.

Now, in the definition of powers (or *exponents*) of numbers, a square root counts as a power of $1/2$, or equivalently 0.5. Looking at this result

$$\sqrt{2} = 2^{1/2} = 2^{0.5} = 1.414$$

many pioneers in algebra came to an interesting conclusion. They realized that if you can use fractional exponents, you can generate any positive number you want as exponents of a single starting number, called a *base*.

For example, using the number 2 as a base, you can find 2 to the exponent 3.5 as

$$2^{3.5} = 2^3 \times 2^{0.5} = 8 \times 1.414 = 11.31$$

and other numbers between 8 and 16 can be generated by picking other fractional exponents (exponents between 0 and 1), it follows that you can generate an approximation of any number as a power of 2. In other words, the equation

$$N = 2^x$$

has a solution x for any positive real number N that you care to pick.

Table 12-1 gives you a little table fragment for use with this argument:

Table 12-1	World's Smallest Log Table
Log Base 2	*Number*
2.0	4.0
2.5	5.66
3.0	8.0
3.5	11.31
4.0	16.0
4.5	22.63

In this case, I've generated a table of powers based on the number two. The table, for historical reasons, is called a *table of logarithms*. In real life, logarithms using as a base the number 10 and the transcendental number *e* are the only ones that we typically use. Base 10 and base *e* correspond to the keys *log* and *ln* on a calculator, respectively.

The reason this is important is that in pre-calculator days, adding numbers was seriously easier than multiplying them. If you make up a table of numbers and their equivalents as powers of some number (the *base*), then you can multiply these numbers just by adding the powers and looking back to the table for the original number that corresponds to your new power. In this table, if you want to multiply 4.0 by 5.66, you look up the logs (2 and 2.5), add them (you get 4.5), and read off the product from the table (4.5 corresponds to 22.63). The slide rules that engineers and scientists used until the appearance of calculators were a sort of logarithm table engraved on sliding strips.

You don't need to know much about logarithm tables anymore — the other day, I saw a Sanyo scientific calculator with log/ln keys in a drug store for $6. Of course, I bought it to add to my collection of about 20 completely unnecessary calculators. (Hey, it's cheaper than collecting 35mm zoom lenses.) As an additional curiosity, even the cheapest plastic slide rules of the 1960s and earlier were so thoroughly discarded when scientific calculators were developed that they are now expensive collectibles, like rare models of Barbie dolls.

Just as a demonstration, I'll show you the meaning of a couple of logs. Using a calculator, I find that

$\log 115 = 2.0607$

$\log 12{,}400 = 4.0934$

Log 100 would be 2 since $100 = 10^2$ (that's essentially the definition for log). Because 115 is a bit more, its log is a bit larger. This result also says that 10 to the 0.0607 power should be the factor 1.15. You can check this yourself on a calculator with a 10^x key — and it checks, all right.

The log of 12,400 is just same thing, scaled up. The log function essentially counts the number of decimal places and as the decimal part calculates another number that's a sort of correction factor to place the number approximately within the next factor of ten.

There's not much more to logarithms than that. They're used in science extensively for dealing with large numbers — after all, the log of a million is six, so they're good for scaling things down to a manageable size. But now that experimenting with logs on a calculator is quick and easy, a lot of the mystery is gone.

Higher Math

In algebra, the translation of a word problem often becomes an expression in x. If the final expression contains just x and numbers, you can always solve it. In fact, if an expression contains just one unknown (x is the unknown here) to a power less than five, you can always find solutions for x. But if the expression ends up containing higher powers of x, sometimes you can solve it and sometimes you can't. In fact, one of the main challenges facing mathematicians for centuries was to determine when you can and when you can't solve an algebraic equation.

In most day-to-day algebra problems, the highest power you encounter is a mere 2. Expressions containing x^2, called *quadratic expressions,* are not only especially common but also especially easy to handle. In fact, you can write a general solution for any quadratic equation. That solution is shown in Figure 12-3.

QUADRATIC FORMULA

What is the quadratic formula?

The quadratic formula is used to solve any equation in the general form

$$a x^2 + b x + c = 0$$

where a, b, and c are all constants. Any such equation is called a quadratic equation.

In mathematical terms, for any quadratic equation $ax^2 + bx + c = 0$ the solutions are provided by the quadratic formula, shown below:

$$x = \frac{-b \pm \sqrt{b^2 - 4ac}}{2a}$$

Figure 12-3:
Quadratic
expressions,
automated.

There's also a pretty complicated solution of equations that contain x^3, although you seldom need them. The reality is that, for finding the values of x that satisfy a general expression like this:

$$ax^5 + bx^4 + cx^3 + dx^2 + ex + f = 0$$

you usually need to have actual numbers plugged in for the *a, b, c,* and so forth, and you use some computer program to find the xs that work by a process that's actually a sort of enlightened trial-and-error method.

Modern computer programs for mathematics typically have an automated method for finding the zeros of polynomials (these expressions in x and powers of x are all called *polynomials,* and you can also have polynomials in several variables). Figure 12-4 shows such a program finding approximate solutions to a nice, big equation.

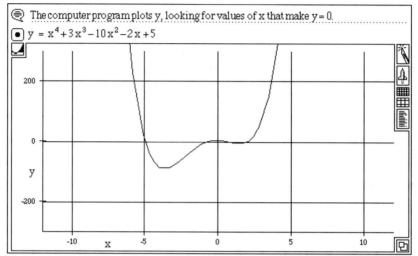

The computer program plots y, looking for values of x that make y = 0.

$$y = x^4 + 3x^3 - 10x^2 - 2x + 5$$

Figure 12-4:
A computer
solution of
an equation.

If you have any interest in seeing how modern computer programs have taken
over traditional topics in algebra, you might want to look at the program
Theorist (from Maple Waterloo Software), available for Windows PCs and
Macintoshes. Just as the program Quicken back in Chapter 1 can do anything
you ever needed done for your checkbook, Theorist can solve every problem
you ever faced in high-school or college math. Although the scientists in *Far
Side* cartoons still work at blackboards, in real offices and labs the work is done
on a computer. Computer algebra has simply revolutionized equation-solving,
probably forever.

Chapter 13

What They Were Trying to Tell You in Geometry

• •

In This Chapter

▶ Geometry before Euclid

▶ The Greek ideal emerges

▶ Descartes changes it all

▶ Graphing makes geometry easier

• •

*E*ven now, a mere 2,200 years or so after Euclid, Euclidean geometry is considered to provide such an excellent mental exercise — often a student's first introduction to the idea of rigorous proof — that it remains part of the high school curriculum. What Euclid introduced into the large body of geometrical results available to the ancient world was the idea of the *axiom*. Instead of accepting the idea that geometry consisted of hundreds of unrelated results, rules-of-thumb, and "guesstimates," Euclid decided to find the smallest number of *axioms*, or fundamental principles, from which all the other rules could be derived.

There is a great deal to be said for this approach, especially for students who will be going on to advanced studies in math or science. It has the unpleasant side effect, however, of producing vast platoons of ex-students who remember only that they hated geometry and can't remember the formula for the area of a circle. Another problem with the Euclidean approach is that it doesn't say much about the way the facts of geometry were discovered. Propositions simply appear from nowhere, waiting to be proven, which is a pretty unnatural state of affairs and, to be honest, not the way mathematicians produce new results in mathematics.

Euclid himself is quoted as saying to one of the Ptolemaic kings of Egypt (he worked in Alexandria) that asking for a simpler approach was pointless because "there is no royal road to geometry." Actually, that depends on where you're headed. There aren't any shortcuts to mastering Euclidean proofs, but learning a few practical geometric facts is no great effort.

B.E. (Before Euclid)

The Babylonians and Egyptians, being rather more practical than the abstraction-minded Greek philosophers, assembled rule books covering most common land-measurement or volume-measurement situations. Learning these rules was part of the education of a scribe. Here are some rules that may be familiar to you.

Area of a rectangle

One of the earliest maps in existence is a little clay block measuring about six inches by three inches (the scribes preferred palm-sized clay tablets) from Mesopotamia. On it, you can clearly see a bunch of standardized riverfront lots going from a fixed boundary down to the Euphrates (see Figure 13-1). Later clay maps (there's a great city map of the city of Nippur from 1500 B.C.) showed the clarity of a modern subway directory. The Sumerians and their successors weren't much for theory, but even they could translate the imprecise boundaries of the land-map plots into ideal rectangles and calculate the area (see Figure 13-2).

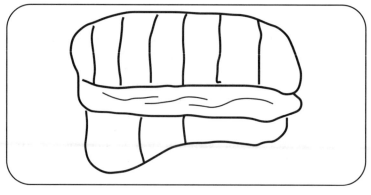

Figure 13-1:
The dawn of geometry: riverfront lots in Mesopotamia.

Sumerian Riverfront lots, depicted on
a little hand-held clay map

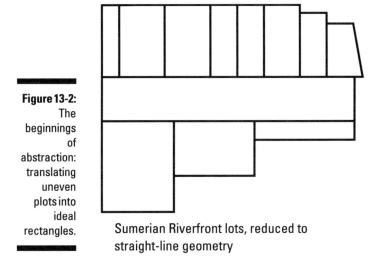

Sumerian Riverfront lots, reduced to
straight-line geometry

Because plots of land were (and are) worth money, real areas were calculated directly on the ground rather than from figures in clay. And the favored measuring tool of the ancient Near East was the knotted rope, as shown in Figure 13-3. We actually have some real examples from the dry climate of Egypt (hemp rope doesn't keep well through the 5,000 years of humidity that constitutes the climate history of Sumeria) and can thereby reconstruct all sorts of geometric tricks of the ancients.

The first and simplest is the area of a rectangle. Take two ropes, line them up at the edges of the rectangle, and count all the little squares, as Figure 13-4 demonstrates. One side of the rectangle is three knots long, and one side is four knots long. If you count the internal squares, you have 12 of them. After you do this for a while, it becomes clear that multiplying the numbers for the lengths of the two sides is exactly the same operation as counting all the squares in the rectangle.

By the way, you can do the same thing if both sides of the rectangle turn out to have the same length, which is the special case of the *square*.

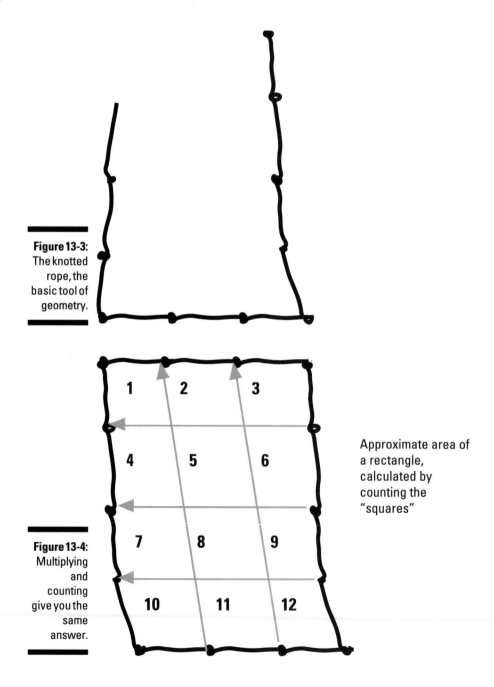

Figure 13-3:
The knotted
rope, the
basic tool of
geometry.

Approximate area of
a rectangle,
calculated by
counting the
"squares"

Figure 13-4:
Multiplying
and
counting
give you the
same
answer.

Area of a bent rectangle

Once you start laying out real riverbank plots on a hot riverbank in a summer in Sumeria in 2750 B.C., what do you think are the odds that the plots will be perfect rectangles? Imperfection is almost guaranteed. The real plots are probably going to look something like those shown in Figure 13-5.

This lot is going to be a special case for area calculations.

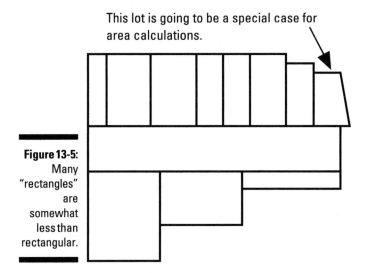

Figure 13-5:
Many "rectangles" are somewhat less than rectangular.

What did the scribes make of this situation? They looked at it and made the following conclusions:

1. We have to multiply the long side by something.

2. If we multiply the long side by the shorter end near the road, the answer will be too small for the correct area.

3. If we multiply the long side by the longer end near the river, we are clearly going to get an area that's too large.

4. What the heck. Propitiate the goddess Inanna and split the difference. We'll take a number in between the values of the two ends and multiply that by the long side. (See Figure 13-6.)

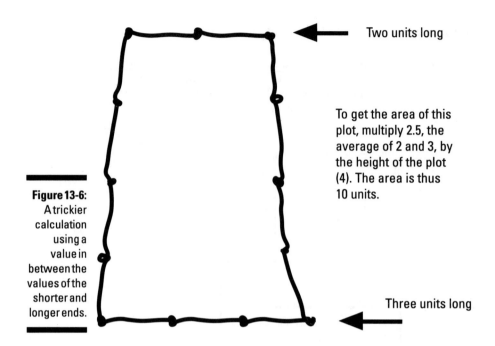

Two units long

To get the area of this plot, multiply 2.5, the average of 2 and 3, by the height of the plot (4). The area is thus 10 units.

Figure 13-6: A trickier calculation using a value in between the values of the shorter and longer ends.

Three units long

When the Greeks got around to developing Greek names for everyone else's geometrical objects 2,000 years later, they named this kind of rectangle a *trapezoid.* Please remember that at the time these calculations were going on in Sumeria, the Greeks were an illiterate Indo-European tribe with nothing to offer their neighbors but an inexhaustible supply of petty skirmishes.

Area of a triangle

This last result is actually fairly sophisticated because it leads the way to other area calculations. Look at the series of "rectangles" in Figure 13-7. The last one has ceased to be a rectangle at all and is now just a triangle. But the rule for calculating the area is still the same as it was in the "bent" rectangle case.

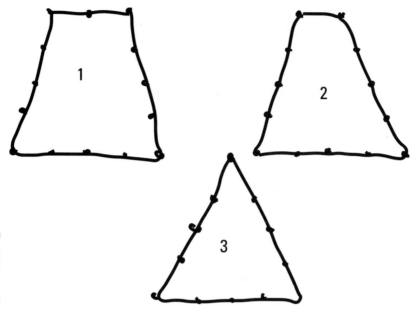

Figure 13-7:
A series of
trapezoids.

When the length of the short end drops to zero (that's the case of the triangle), the rule becomes

$$\text{area} = \text{length} \times \frac{(\text{shortside} + \text{longside})}{2}$$

$$\text{area} = \text{length} \times \frac{(\text{side that's left})}{2}$$

In standard notation, this is the rule

$$A = \frac{(a \times b)}{2}$$

where the letters stand for the lengths in the diagram in Figure 13-8. This is the standard formula for the area of a triangle, usually derived by drawing a triangle inside a rectangle (see Figure 13-9) and comparing the areas of the different sections.

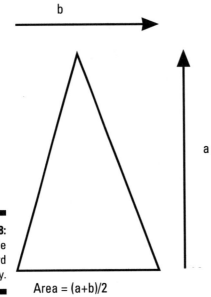

b

a

Figure 13-8:
The triangle
in standard
geometry.

Area = (a+b)/2

b

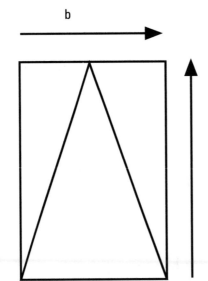

Area of the
rectangle is ab,
while the area of
the triangle is just
half of that.

a

Figure 13-9:
The usual
derivation of
the triangle
area
formula.

Area = (ab)/2

Area of a circle

Equipped with the formulas for the area of a rectangle and a triangle, geometers around 2000 B.C. were ready to measure the area of any sort of geometrical object, no matter how oddly shaped. Doing so is simply a matter of breaking up the object into rectangles and triangles, working out the area for all these little bits, and adding up the areas of the bits. As a practical consideration, a city official could work out the exact taxes on any plot of land, undoubtedly a comforting fact for the hard-pressed Sumerian peasant. No wonder these people invented beer.

But the method of calculating areas by looking at areas of rectangles or triangles is surprisingly powerful. Look at the two possibilities in Figure 13-10.

- ✔ If you draw a circle inside a square (10A), you can conclude that the area of a circle has to be smaller than four times the area of one of the little subsquares in the figure. If the area of a little subsquare is one, then the area of the circle is less than four.

- ✔ If you draw a square inside this same circle (10B), you conclude that the area of the circle has to be larger than the area of the four little subsquares. The little subsquares in this part of the figure have areas of $1/2$ area unit, compared to the subsquares in 10A.

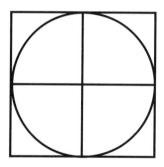

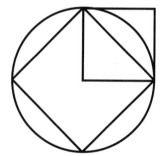

10A: The area of this circle is less than 4, if the squares have area 1.

10B: The area of this circle is more than 2 (the circle surrounds four half-squares).

Figure 13-10: Estimating the area of a circle by using squares.

This square is a unit of area for this calculation.

✔ In these units, the area of the circle is somewhere between four and two since those are the areas of the outside square set and the inside square set. In this diagram, the side of one of the squares in 10A corresponds to the radius of the circle (the radius goes from the middle of the circle to the edge). Following traditional Mesopotamian logic, looking at the numbers four and two, you would make an offering to the memory of Gilgamesh and say that the area of a circle is about *three* times the square of the radius. In modern terms, the rule would be

$A = 3 \times r^2$

If you remember anything from geometry, you may remember that nothing to do with circles ever has to do with the number 3. It always has to do with the mystery number π, which is now known to be 3.141592653..., on and on with seemingly random digits.

Three is a nice, clean number, but even the ancients knew that they weren't getting away this easy. Using squares to try to cover a circle is a pretty sloppy procedure. Look at the diagram in Figure 13-11 for an improved covering, also known in ancient Mesopotamia and Egypt. By arranging triangles into *hexagons* (that nice, six-sided, turn-of-the-century bathroom tile shape is a hexagon) and taking area averages of the outside hexagon and the inside hexagon, the value for the area of the circle starts to look approximately like 3.15 times the radius of the circle squared. 3.15 is in fact a pretty good guess at the real value of π. It means that all these ancient calculations on circles were good to within 1 percent, which is probably better than the original accuracy of a knotted-rope measurement, anyway.

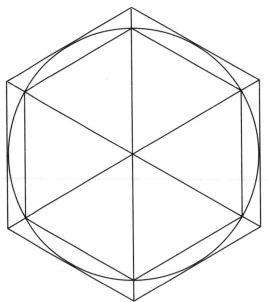

These hexagons cover the circle more exactly than squares do.

Figure 13-11: A better fit for a circle's area.

Both the Babylonians and the Egyptians had a distinct preference for expressing fractions only in the form

fraction = 1/something

The fractional part of this number looked like about $1/7$, so the number $3 + 1/7$ came to be widely accepted in the ancient world as a good working value for π (although nobody called it π, a Greek letter, until the Greeks took over the subject). This value was resoundingly good enough for two reasons:

- ✔ If you are measuring areas outdoors with knotted ropes, you probably only need to get within one percent of the correct area.
- ✔ Nobody in the ancient world had circular fields, so the whole exercise was a bit theoretical anyway.

And now, turn up the volume

So there are ancient formulas for areas of rectangles, triangles, and circles. One last useful quantity to estimate is *volume,* since temple tributes in the ancient world were often stated in terms of a certain volume of grain. In fact, your typical tax problem in the ancient world was that the local authorities kept a record of the area of land that you had under cultivation, and had worked out an estimate of the volume of grain that you were likely to produce from it. And also, of course, the officials estimated how much you owed them for their services in keeping order and regulating rainfall and river floods with prayer.

As you might expect from a down-to-earth bunch like the inhabitants of ancient alluvial valleys, the first rules for volume were probably worked out by using little models. After all, these people were up to their eyeballs in mud bricks, objects that lead directly to speculation about volume. If you think about it for a minute, you will realize that if you know how many bricks went into a solid brick temple platform, you automatically know the volume of the platform.

For two simple volume objects, the brick and the cylinder, the ancients realized that you just multiply an area times a height. Probably by direct experimentation (you can do the same thing yourself with a big box of sugar cubes), they apparently also worked out the volume of a cone (see Figure 13-12).

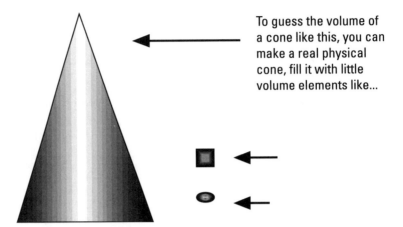

To guess the volume of a cone like this, you can make a real physical cone, fill it with little volume elements like...

Figure 13-12:
Volume
formulas.

...and see what makes a reasonable formula. One likely guess is Volume = area of circle at bottom x height x factor. The factor turns out (by ancient experiment or by calculus now) to be 1/3.

With the simple repertoire of area and volume formulas summarized in Figure 13-13 and a certain willingness to break eggs when making an omelet (you didn't want to get on the wrong side of an Egyptian tax collector, who was supposed to make up any shortfalls from his own account), it's possible to keep together a civilization that lasts many centuries. There was little philosophical content to the scribes' rule books, but they could tell you how many bricks you would need to beef up the city walls against the invaders from the desert.

Philosophy Enters the Picture

Greek contact with the knowledge of the ancient world changed everything. You may have noticed that the first part of this chapter was probably easier to follow than Chapter 9 of your high school geometry book. That's because the Sumerians, Babylonians, and Egyptians just wanted to get their measurements right in a practical sense. By the time of Greek ascendancy in the eastern Mediterranean, the most useful rules of geometry were already 2,000 years old.

The Greeks were interested in the Why of things, in a close examination of fundamentals, and typically delighted in hair-splitting, exasperating argument. Culturally, they were what you call a Hard Act to Follow in several senses. It's hard to imagine anyone else in the ancient world arguing about Good or Virtue — elsewhere, Good simply meant having enough to eat for the first few thousand years of recorded civilization.

Area of a rectangle = a x b

 Volume of a brick = a x b x c

 Area of a triangle = (a x b)/2

 Area of a circle = pi x r x r = pi x r^2

 Volume of a sphere = (pi x r^3) x 4/3

 Volume of a cylinder = pi x r^2 x h

Figure 13-13:
Simple
formulas.

 Volume of a cone = (pi x r^2 x h)/3

I don't propose in one chapter in a book called *Everyday Math For Dummies* to review to any extent the idea of axiom-based proof, mainly because that's not everyday math (when was the last time you needed to prove a theorem in daily life?). But I hope to suggest how Greek civilization transformed mathematics from a down-to-earth counting enterprise into the abstract activity it subsequently became. I have been encouraged to do this by a committee of high school math teachers, who figure that I can spare a few pages out of this book for general cultural literacy. If you can't stand this stuff, reread Chapter 11.

Pythagoras

The Greek philosopher and mystic Pythagoras (who lived in southern Italy around 550 B.C.) is best known in geometry textbooks for the Pythagorean theorem (quoted exactly in the movie *The Wizard of Oz* by the Tin Man as he's awarded his diploma). Actually, a form of the theorem was known to earlier civilizations with their funky old knotted ropes (see Figure 13-14), but only for particular triangles.

There's reason to believe that the main Pythagorean result about right triangles (triangles in which one corner is a 90-degree angle, or *right* angle) was known for a handful of special right triangles. It's characteristic of Greek intellectual effort that Pythagoras (or his followers, or someone in the neighborhood) sought to generalize the old knotted-rope result to all right triangles. The theorem, with a diagram used in a standard proof, is shown in Figure 13-15.

It was known in the earliest civilizations that when the sides of a triangle fit the ratio 3, 4, 5, the triangle is a right triangle.

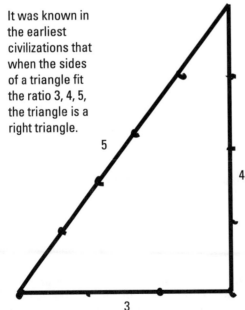

Figure 13-14:
A 3-4-5
triangle in
rope
measurement.

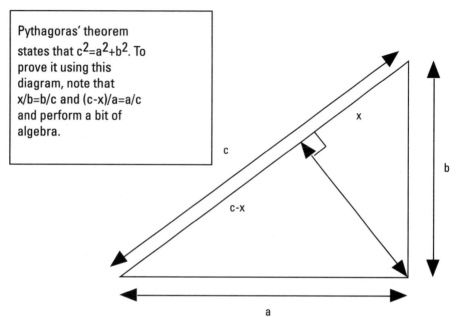

Pythagoras' theorem states that $c^2 = a^2 + b^2$. To prove it using this diagram, note that $x/b = b/c$ and $(c-x)/a = a/c$ and perform a bit of algebra.

Figure 13-15:
The Pythagorean theorem.

There's a nifty proof of the same result from China (around 40 A.D.). Einstein also discovered a proof of his own when he was a teenager. I'm particularly impressed that one proof was also discovered by U.S. President James Garfield. I confidently expect that we will never again see anyone at that political level who can balance his or her own checkbook, much less come up with an original proof of a classic math result.

Pythagoras and his followers, in addition to the highly useful and famous theorem, also had an extensive cult of magical properties of numbers. Triangular numbers (see Figure 13-16) were thought to be better than others, the tuning of musical intervals on stringed instruments followed a complicated set of beliefs about harmonic intervals, and in general the rules of arithmetic were invested with all sorts of moral and philosophical significance. Because they were written down in manuscripts that survived to Renaissance times, these beliefs promoted lots of mathematical mysticism as late as the 1800s in some circles. The Greeks weren't all that logical.

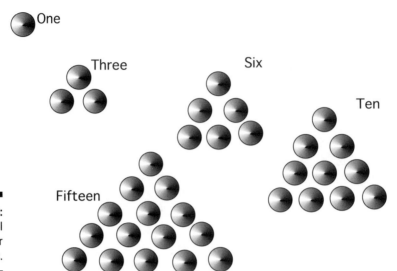

Figure 13-16:
The mystical triangular numbers.

Zeno

Like Pythagoras, Zeno came from the Greek colonies in Italy (around 480 B.C.) and made it to Athens in time for the main events of the Golden Age when Athens was the leading city of the Greek world. He contributed some interesting paradoxes that succeeded in muddying the intellectual waters on questions of motion and time until the time of Newton. You can only get so far by thinking about things and denying common sense, although this originally Greek pastime is alive and well in many locations today. Find yourself a coffee house near a university if you don't believe me.

In one paradox, Zeno claimed that, basically, you couldn't get where you were going. First you would get halfway there, then you would get half of the rest of the way, then half of the rest of that, and so forth. This process of covering the next little bit of distance clearly would take forever (see Figure 13-17). In a related paradox, Zeno argued that the mighty Achilles could never overtake a tortoise with a head start, because in the time it takes Achilles to get to the tortoise, the tortoise has moved on, so there's another interval to close. Zeno's paradoxes were meant to suggest that you couldn't rely on the evidence of your senses. Even if you could catch up to a tortoise, Zeno could prove that you couldn't. So there.

The resolution of these issues is that it doesn't take an infinite amount of time for all these successively smaller gaps to close. Later in this chapter, in looking at Descartes' contribution to geometry, I'll revisit Zeno's paradox, mainly because I have always found it annoying that it was taken seriously for centuries. What was the matter with these people?!

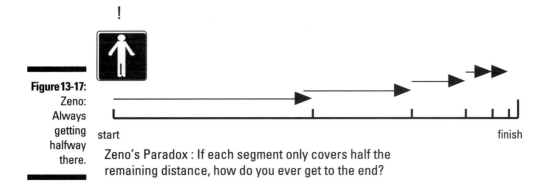

Figure 13-17: Zeno: Always getting halfway there.

start   finish

Zeno's Paradox : If each segment only covers half the remaining distance, how do you ever get to the end?

Euclid

Euclid, who lived around 300 B.C. in a Greek world that had been transformed by Alexander the Great, wrote 13 books called the *Elements,* the first few books of which still constitute, in somewhat watered-down form, the first half of high school plane geometry. No one has the slightest idea how much of the material originated with Euclid — we have only a small percentage of all the writings of the ancient world, and for all anyone knows, Euclid's work could be a copy of the works of other Alexandrian mathematicians.

In Euclidean geometry, you simply start with a few undefined geometric elements and a small set of axioms (see Figure 13-18) and proceed to derive all the geometric results of the previous few thousand years.

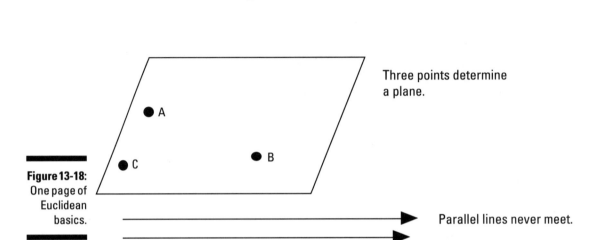

This is a point (primitive or undefined concept).

Two points determine a line.

A   B

Three points determine a plane.

A

C   B

Figure 13-18:
One page of
Euclidean
basics.

Parallel lines never meet.

The emphasis of this effort is quite different from that of earlier geometry. For one thing, nothing is measured. Constructions are made using only a straight-edge (for drawing straight lines) and a compass (for drawing circles). You aren't allowed to make absolute measurements on a diagram — you might, for example, prove that two angles in a diagram are equal (see Figure 13-19), but you never actually measure the angles. Similarly, measuring areas or distances is "cheating" in this form of geometry — questions must be settled by argument rather than simple recourse to an experiment on paper or with materials.

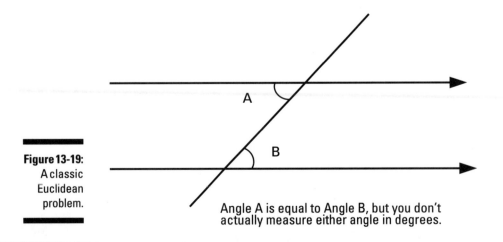

A

B

Figure 13-19:
A classic
Euclidean
problem.

Angle A is equal to Angle B, but you don't actually measure either angle in degrees.

You can get the flavor of this enterprise, and something of its sociological motivation, by quickly examining this proposition:

The ratios of the lengths of corresponding sides of equiangular right-angled triangles are equal.

This statement has lots of practical consequences, as shown in Figure 13-20, but is sufficiently annoying to prove that it was known for centuries as the *Pons Asinorum,* in Latin the "Bridge of Donkeys," meaning that you're on the wrong side of the bridge (back with the donkeys) if you can't prove it. I believe that this particular bit of terminology calls into high contrast the rather unpleasant "proving you're smarter than the other guy" aspect of much of mathematics education, and a long-standing aspect it is, too.

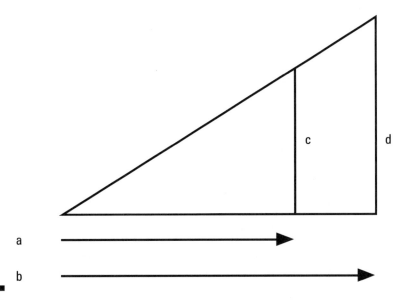

Figure 13-20:
Using a theorem of Euclid.

Theorem: Show that c/a = d/b, where the letters stand for lengths of the indicated line segments.

In high school classes, it is seldom remarked that the ambitious 19th-century program to *axiomatize,* or make rigorous, all of mathematics, led to an interesting discovery. The discovery was that this can't be done (see Chapter 25). Greek mathematicians may have thought that every important result in geometry could be proven from a small set of axioms, but it has subsequently developed that this belief is wrong. This certainly doesn't mean that the idea of proof is outdated or that learning to prove theorems is a waste of time, but it does mean that the real world of mathematics has turned out to be a lot stranger than the Greeks imagined.

The New World of Geometry

The first real innovation in geometry in close to 2,000 years occurred when the brilliant French philosopher René Descartes began to explore the idea of assigning numbers to points in a plane. Even if you aren't familiar with this idea, you use essentially the same principle every time you look up Whatzis St. on a city map and find it in D-6. In fact, a quick inspection of Figure 13-21 may convince you that Descartes' big idea was sitting there waiting to be discovered in Egypt.

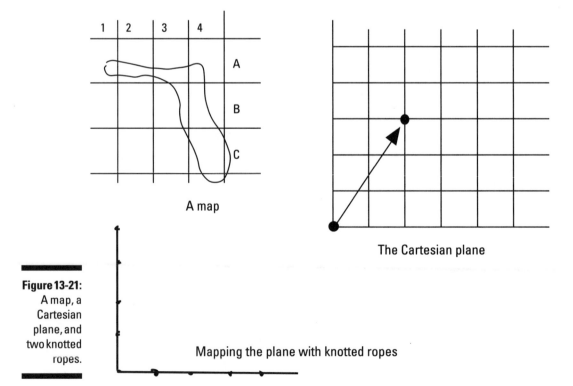

A map

The Cartesian plane

Mapping the plane with knotted ropes

Figure 13-21:
A map, a Cartesian plane, and two knotted ropes.

When Descartes, loafing around in his bed all day in exile in Amsterdam (around 1630) developed the subject of analytical geometry, he really founded a whole family of subjects that form the basis of applied mathematics today. By *analytical geometry*, mathematicians really just mean geometry with numbers in it — the geometry of graphs.

In the geometry of Euclid, there are essentially no numbers — the notion of quantity is expressed in terms of ratios. That's not how you design airplane wings. In Cartesian geometry, by contrast, numbers are everywhere, even to the extent that you can use numbers to define space with more dimensions than you can actually see. For this book, I'll just use examples from the plane, but everything that I say in this section applies to seven-dimensional space just as well.

Distance in a plane

Here are a few conventions in Descartes' geometry, called *Cartesian* geometry. The "up" direction in a Cartesian plane is usually called the *y*-direction, and sideways is the *x*-direction. Every point in the plane is identified by its *x*-position (called the *x* coordinate) and its *y*-position (called the *y* coordinate). The point (3,4), for example, means the point *x* = 3 and *y* = 4 — there it is in Figure 13-22 — you go sideways three units and then up four units to locate it.

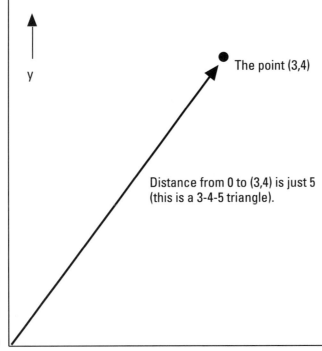

y

The point (3,4)

Distance from 0 to (3,4) is just 5
(this is a 3-4-5 triangle).

Figure 13-22:
The point
(3,4) and its
distance
from 0.

x

If you look at the figure again, you may see the outlines of your friend the Pythagorean theorem. In fact, you use this theorem to compute distance in the plane. Between every two points is a line that turns out to be the long side (*hypotenuse*) of a right triangle. Because it's easy to calculate the other two sides of the triangle, you can calculate the long side as well; the long side is just the distance between the two points (see Figure 13-23).

A line or two

Now, since I'm calling the first number in a pair the *x* coordinate and the second number the *y* coordinate, every time I write down some relation between *x* and *y*, I have defined something that turns up in the Cartesian plane as a piece of geometry. Consider two examples.

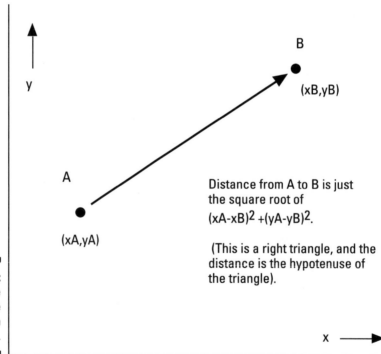

Distance from A to B is just the square root of $(xA-xB)^2 + (yA-yB)^2$.

(This is a right triangle, and the distance is the hypotenuse of the triangle).

Figure 13-23:
The distance between two points.

First, what is the meaning of this expression?

$$y = 2x$$

To find out the meaning, you pick a bunch of x-numbers, find the corresponding y-numbers, and then locate all these (x, y) pairs on the plane. This process should be pretty simple. Pick 1 for the first x. The expression says that the y that goes with this x is just $2x$, which is 2. So one of the points is (1,2). Pick 3 for the next x, and the expression says that 6 is the corresponding y. That point is (3,6). In this way, you just make a short list of points and see where they fall on the plane. Check it out in Figure 13-24.

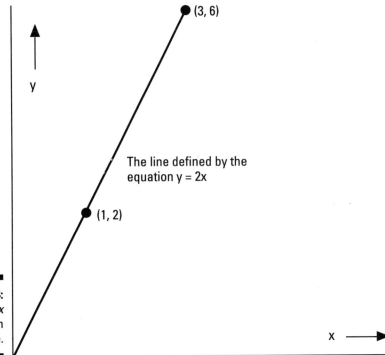

The line defined by the equation y = 2x

Figure 13-24: $y = 2x$ plotted in the plane.

There. That's not so bad. You have now *plotted a straight line in the Cartesian plane*. Didn't hurt a bit. Emboldened by this success, try just one more modest advance.

Consider this expression:

$$y = x - 2$$

What happens here? For $x = 0$, the rule says that $y = -2$, giving you the point $(0,-2)$. For $x = 8$, the rule says that $y = 6$, giving you the point $(8,6)$. Collecting a few more points, you can make the plot in Figure 13-25. Note that the rule doesn't care if you put negative numbers like $x = -17.1$ into the expression — it's all the same from the point of view of plotting.

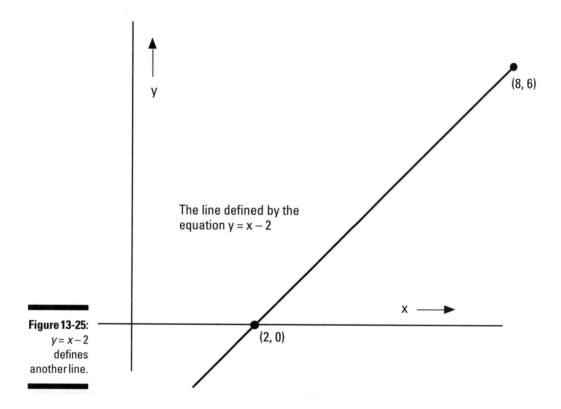

The line defined by the equation $y = x - 2$

$(8, 6)$

$(2, 0)$

y

x

Figure 13-25:
$y = x - 2$
defines
another line.

It happens that every straight line in the plane is defined by an expression relating y and x this way (in a little bit, you'll see what happens when you have x^2 or something instead of plain x). All the constructions of Euclidean geometry involving straight lines can be discussed in terms of these x-y equations. Not only does this make all sorts of proofs almost straightforward, but it allows a nice extension to curves as well, as you'll see in the following section.

A circular argument

In Figure 13-26, I have drawn a circle around the point (0,0). How would you describe this in x-y terms? Well, the definition of a circle is that every point is the same distance from the center. A few paragraphs back, I also gave a definition of distance.

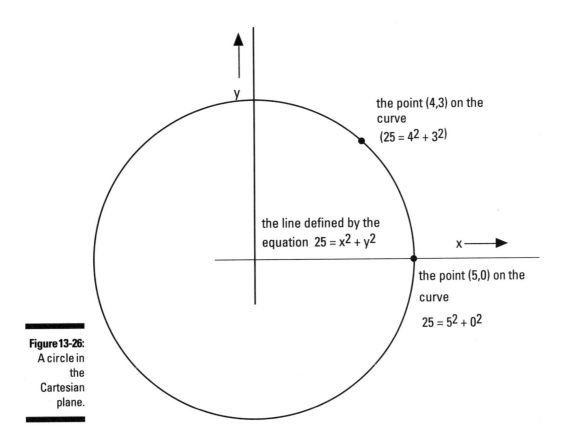

the point (4,3) on the curve

$(25 = 4^2 + 3^2)$

the line defined by the equation $25 = x^2 + y^2$

x

the point (5,0) on the curve

$25 = 5^2 + 0^2$

Figure 13-26:
A circle in the Cartesian plane.

In this case, the center is zero and the distance is always five units, so I'm going to try cooking up an expression this way:

distance from $0 = 5$

Using the distance formula in Figure 13-23, you have

distance $= \sqrt{(x^2 + y^2)} = 5$

or

$x^2 + y^2 = 25$

You can get out a calculator and show for yourself that points (x, y) that satisfy this expression all lie at exactly five units from zero in the plane.

Graphing

One of Descartes' great accomplishments in founding analytic geometry was to greatly reduce the strain on your own brain that some aspects of geometry induce. In the world of Euclid, the subject of conic sections was fraught with notable difficulties, for example. A *conic section* (ellipse, parabola, or hyperbola) is a curve that you get by slicing a cone with a plane.

In the new world of finding points from a rule and plotting them, you can prove all sorts of Euclidean theorems by performing a few simple algebraic operations on the expressions for the curves. Not much of this subject counts as everyday math, but Figure 13-27 shows what these curves look like for your general background. There's nothing in these plots that you couldn't do yourself just by making up the sets of points dictated by the expressions and then dumping those points onto the plane. Relatively inexpensive graphing calculators can do all this work for you, anyway. There is still no royal road to geometry, but access to the basic facts of geometry is certainly no longer the formal challenge to an elite that it was in 300 B.C.

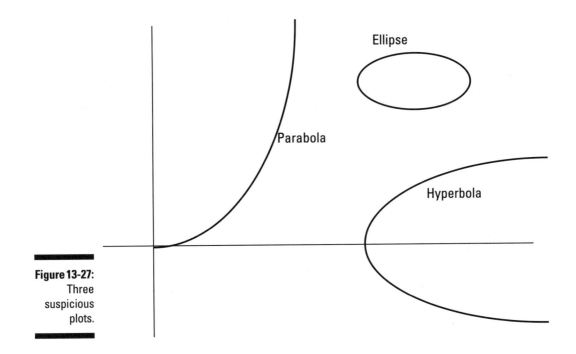

Figure 13-27:
Three
suspicious
plots.

Chapter 14

What They Were Trying to Tell You in Trigonometry

1 expect you would have noticed by now that a lot of education has to do with setting up social distinctions. In a good high school, many of the kids who are going to college take Latin; the kids who have been consigned to the vocational programs probably take one year of Spanish in a course designed to impart very little fluency. In a country that thinks of itself as a democracy (and has certainly done a good job, at least for a big country), there's an amazingly detailed social pecking order, not just among American schools but also among programs within schools.

Trigonometry, often in the form of a course called Advanced Algebra, is one of those courses that typically distinguishes college-bound from non-college-bound students in high school. If you didn't take it, or have forgotten nearly everything, I have two mutually contradictory bits of good news for you:

✔ You're not missing much, at least in practical terms. If you had to pick one math topic to master, compound interest outweighs trig by a lot.

✔ All the hard parts of trigonometry have been made fairly simple with the introduction of the scientific calculator.

This chapter should probably be titled "Cultural Literacy about Angles" because it discusses lots of topics but doesn't pose much problem-solving. For example, you wouldn't think that you'd find stuff about radio in a trigonometry chapter, but that's just one good example of the way all sorts of math topics are interrelated.

In the Land of Sines and Cosines

I once took a chemistry course in which the professor, who had been discussing the way scientists synthesized chemicals in the last century, looked up at the class as if he had surprised himself and remarked, "Ya know, those old guys weren't so dumb." That's even more true of mathematics, where an Egyptian scribe could probably stomp the average high school senior on a suitably translated SAT test. Hence this bit of historical review.

Long ago and far away: ancient Egyptian measurement

In Greek geometry, you spend lots of time proving that some angle is equal to another angle, but you never have an actual measurement, unless of course it's a right angle.

Egyptian geometry places a big emphasis on having a number for the angle under discussion. In terms of how many blocks of limestone you need, it makes a big difference whether a pyramid is going to be slightly flatter or slightly steeper. (*Slightly* is the right word here, too, because all pyramids use nearly the same angles.) The Egyptian measurement of angle is the *seked* (see Figure 14-1), and the sides of pyramids have angles between $5\frac{1}{4}$ and $5\frac{1}{2}$ *seked*.

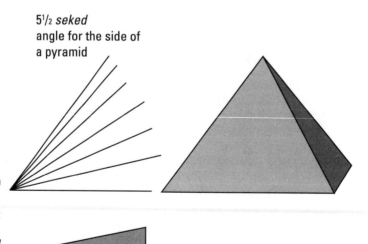

$5\frac{1}{2}$ *seked*
angle for the side of
a pyramid

Figure 14-1:
The great
pyramid,
a $5\frac{1}{2}$-*seked*
creation.

A one-*seked* wedge block

The practical Egyptians probably liked the *seked* because it's a good angle for a wedge or a ramp. In this civilization, in contrast, we divide the angles of a whole circle into 360 degrees. We inherited this number of degrees from the Babylonians, who were fond of multiples of 12 and 20. Like the early astronomers of Mexico, the Babylonians felt that the year *should* have had 360 days because 360 is such a cool number. Practically every early civilization had the year organized into 360 days that were okay and 5 bad-luck days stashed at the end. With 360 degrees in a circle, you have the tableau of appealing and popular angles shown in Figure 14-2.

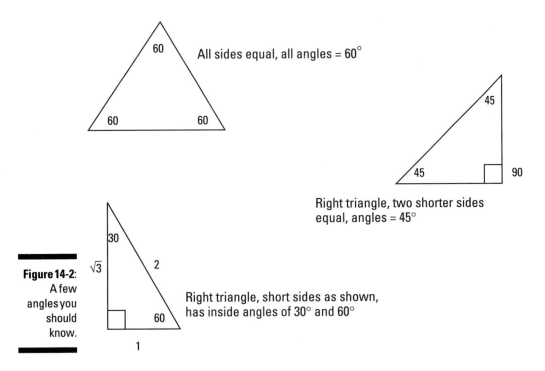

Figure 14-2: A few angles you should know.

Right triangle, two shorter sides equal, angles = 45°

Right triangle, short sides as shown, has inside angles of 30° and 60°

Triangles in a circle

If you have a little one-*seked* right triangle, you can guess that a very big one-*seked* right triangle will be in proportion to it. (See Figure 14-3.) In 2000 B.C., the Egyptians knew that they could use this trick to estimate distances they couldn't actually measure, and that's just what they did (see Figure 14-4).

The height of the wedge block is one unit, and it's five units long. If the palm tree is 25 units away, then it must be 5 units high, since to keep everything in scale we must have

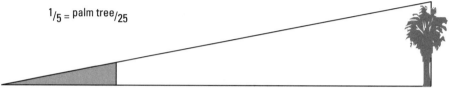

$$1/5 = \text{palm tree}/25$$

Figure 14-3:
The Egyptian triangle gimmick.

a one-*seked* wedge block

If the surveyor knows the distance between the palm trees (by looking straight across the river), he can tell the distance to the trees by lining up a 45° trangle as shown.

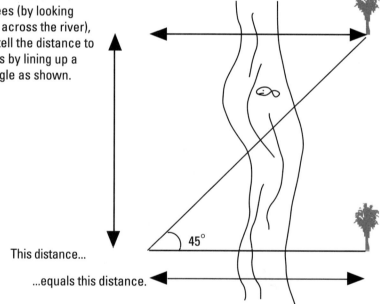

Figure 14-4:
The triangle gimmick in surveying practice.

This distance...

...equals this distance.

The question facing you is this: Can you use this proportional-triangles trick for any angle? The answer provided by a series of mathematicians in the 17th century is: You bet you can!

Here's how the process was generalized. If you draw right triangles inside a circle as in Figure 14-5, you make a whole series of triangles for which you know two things:

✔ The radius of the circle is fixed at one.

✔ s^2 and c^2 together have to add up to one, because of Pythagoras' theorem about right triangles.

Here I'm using my own names for the sides of the triangles — straight up is s and crossways is c. The real names for these triangle sides are sine (s) and cosine (c), abbreviated as *sin* and *cos*. Out of deference to the Greeks, the angle in this situation is usually called theta (Θ), but I'm going to call the angle a because I am a plain-speaking person from the Midwest.

This circle has radius = 1.

For any angle a, the definitions are

$sin\ a = s$,

where s is the length of the upright line segment, and

$cos\ a = c$

where c is the length of the sideways line segment.

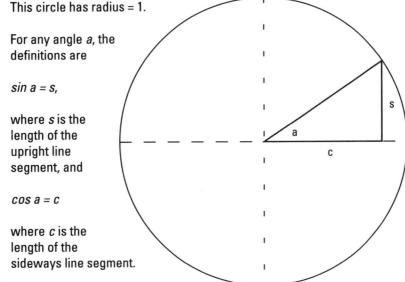

Figure 14-5:
The triangles in a circle define sines and cosines.

Sine, cosine, and surveying

The number for sine depends on the angle a. Table 14-1 shows a little table of sines. Because the sine depends on the size of the angle, you say that sine is a *function* of angle a and write an expression like this:

$0.5 = sin(a)$ for $a = 30$ degrees

You change a, and sin(a) changes. That's the meaning of a function. If you know a, you can look up sin(a).

Table 14-1	A Little Sine Table
sin 10°	0.17
sin 30°	0.50
sin 45°	0.71
sin 60°	0.87
sin 90°	1.00
sin 180°	0.00

Now, how does any of this amount to an improvement on the old Egyptian Triangle Trick for surveying? Here's how: In trigonometry, if you have the length of just one side of a right triangle and the length of a side of a *live* angle (any angle but the 90-degree angle), you can calculate the other sides. In the first version of Trigonometry Against the Pyramids, you get a diagram like the one in Figure 14-6.

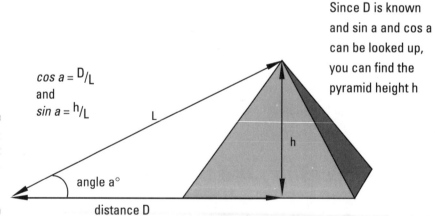

Since D is known and sin a and cos a can be looked up, you can find the pyramid height h

$$cos\ a = {}^D/_L$$
and
$$sin\ a = {}^h/_L$$

L

h

angle a°

distance D

Figure 14-6:
The great pyramid, now a trig problem.

But trigonometry has even fancier tricks than that. Any triangle can be decomposed into two right triangles, which means that it's possible to cook up formulas to cover all sorts of examples in surveying (see Figure 14-7), even if it's inconvenient to formulate the problem in terms of right triangles. And that's just what you see surveyors doing—measuring distances and angles to make maps.

The sides of this triangle have lengths A, B, and C. If you know two of them and can measure the angle a, then you can find the length of the third side from the rule

$$A^2 = B^2 + C^2 - 2BC\cos a$$

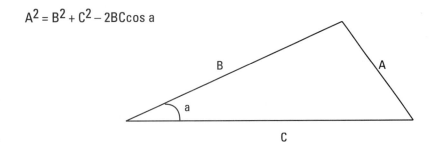

Figure 14-7:
Trickier
problems,
trickier trig.

Angles on the Earth

The previous section talked about *plane* trigonometry, measurements taken on a flat surface, or at least a surface that's flat enough that curvature can be ignored. But angles can just as easily be defined on a *sphere*. As this is being written, my approximate location is 123 degrees west and 39 degrees north. We all live on the surface of a sphere, and the radius is essentially constant — even Mount Everest, at 5 miles high, amounts to a tiny part-per-thousand surface blemish on this smooth globe with a radius of 4,000 miles. That means that our location in three-dimensional space can be closely specified by just two angles, as shown in Figure 14-8. So how did people find these angles out on the featureless sea?

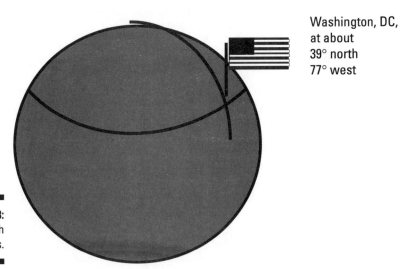

Washington, DC,
at about
39° north
77° west

Figure 14-8:
Home, with
angles.

By late medieval times, mariners had figured out that they could determine how far north or south they were on the planet by sighting the sun with an instrument that measured the angle of the sun above the horizon at noon. (Despite what you may have been told, by 1350 or so, navigators in Europe and the Arab world not only knew the earth was round but they also had a pretty good idea of its diameter.) How does the sun's angle relate to north and south?

Up, down . . .

Here's how. Picture yourself on a sunny day in March, standing around doing nothing near the North Pole (remember, there aren't even penguins to play with). At midday, you sight the sun and you're looking practically straight sideways. Now picture yourself lounging around Quito, Ecuador (why do you think they call it Ecuador?), with a Pisco Sour in your hand, sighting the sun at noon from the front of a fake-Incan gift store. That's right, the sun is directly overhead.

So anywhere in between, from the equator to the North Pole, you would find an angle somewhere between straight up (90 degrees) and straight sideways (0 degrees). That gives you a *latitude* (the number of degrees north or south of the equator). The instrument used for measuring the angle looks just like a protractor from high school geometry class, and lovely brass examples taken from real ships can be found in museums everywhere.

As a curiosity, you might ask what one angle is worth in terms of distance. The world has a diameter of about 8,000 miles, so according to the geometry rules in Chapter 13, you just have a circumference of π times the diameter, which is $3.14 \times 8,000 = 25,120$ miles. (The circumference of the world is the distance around the world at the middle.) Since there are 360 degrees in a circle around the world, one degree of latitude is worth about 70 miles (you divide 25,120 by 360).

All around

Finding the north/south location isn't the difficult problem. The problem is measuring the sideways direction, called *longitude*. You see, the planet is spinning, a great inconvenience in making some directional sightings. Plenty of stars provide reference points for you, but you're spinning around in the middle of them. If you had a star map and a way to tell time, you could perform your sightings at the same point every night.

Alternatively, if you had a time reference for one fixed point on the earth's surface and also your own clock, you could compare the time of sunrise in your location with the time of sunrise at the fixed point (you look it up in a table provided by navigation authorities). This pretty much tells you what your *time zone* is.

All these time zones are referenced to Greenwich, England, and longitude zero is assigned to a line running through the Royal Naval Observatory at Greenwich. (Remember, latitude zero is the equator.) Not coincidentally, the British Admiralty commissioned the development of clocks that could hold time accurately on board a ship ploughing through 10,000 miles of roaring Pacific storms. These clocks (called *chronometers*) made it possible to provide great accuracy in the angle of longitude, so little specks in the vast ocean, such as Niue or Norfolk Island (look 'em up!), could be placed definitively on a navigational chart.

Man-made stars

After centuries of cleverness in navigation, the whole business has now been cleaned up electronically for the benefit of absolute dunces. One of the main motivations of the first satellite builders was to launch a satellite-based navigation system that could give accurate locations anywhere, at any time.

The first systems to interpret the satellite navigational signals were the size of a refrigerator and cost hundreds of thousands of dollars — the appropriate vehicle for using these systems was an aircraft carrier. Now, years later, a calculator-sized navigational system that automatically reads out latitude and longitude with enough accuracy to place you within 20 feet or so costs about $300. I have a friend who bought one for his speedboat, used only for water-skiing on Lake Sonoma. You can't get lost on this lake — you can see the dock from any point, for example. But it was so inexpensive (from a hardware store, too — not even a proper marine supply store) that he couldn't resist. You can buy one and track your trip to Grandma's for Thanksgiving in angular coordinates on the planet from start to finish.

Sines and Waves

And when you get back from Grandma's, trigonometry enters your home in a few interesting ways — literally, in the form of sound waves, radio waves, and all sorts of other phenomena.

Simple waves

Figure 14-9 is a plot of $\sin(a)$ as a runs through all the angles from 0 to 360 degrees. At 360 degrees, of course, the angle is right back where it started, so the next part of the plot looks just the same (see Figure 14-10). What you're looking at in this plot is the simplest kind of wave.

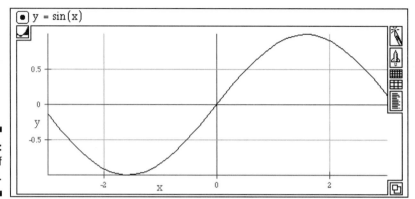

Figure 14-9:
One cycle of
a sine wave.

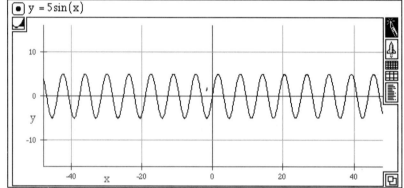

Figure 14-10:
A wave
extending
for many
cycles.

Pinggggg: sound waves

If you strike a tuning fork, it sends a sound pressure pattern out into the room that matches this sort of wave. A *sine wave,* as it's called, corresponds to a pure single tone. If you actually measure the tone with a special microphone, the height of the peaks shows how loud the tone is, and the spacing between the peaks gives the wavelength. Wide spacing means a low note; narrow spacing means a high-pitched tone (see Figure 14-11). In cycles per second (the measurement of frequency now usually called *hertz,* abbreviated Hz), you can hear tones down to about 30 Hz or so and up to perhaps 12,000 Hz. Higher tones, for example, 20,000 Hz, are in the range of dog whistles — you can't hear tones that high.

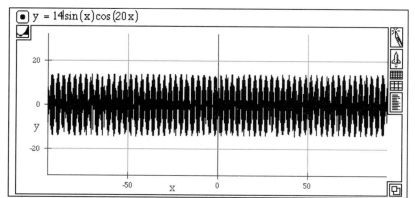

$y = 14\sin(x)\cos(20x)$

So you can have waves of different heights (heights are called *amplitudes* in wavespeak) and different frequencies. The feature of these waves that gives sound its variety is the rule that waves can simply be added to each other. Figure 14-12 shows a whole collection of waves added to each other. Any sound can be represented as a set of waves added to each other to produce a waveform. Complex sounds like speech have very complicated waveforms — that's one reason why computer recognition of speech is such a difficult endeavor. The sounds of real musical instruments, too, are never pure tones but complicated mixtures. That's why it took years of research to make synthesized sounds that sound reasonably close to the sounds of a real piano.

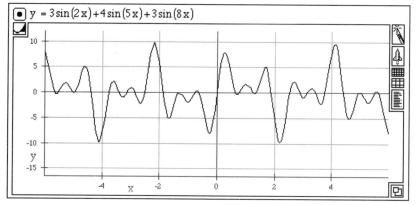

$y = 3\sin(2x) + 4\sin(5x) + 3\sin(8x)$

It also happens that cosines have their own waveforms, which look exactly like sine waves except that their *zero points* are shifted 90 degrees. You can add these to sine waves, and for that matter multiply sine and cosine waves, giving fantastically complex and interesting wave shapes. And every one of these shapes corresponds to a sound that you can actually hear. A music synthesizer just adds up waves electronically and then shapes the amplitude of the resulting waveform.

At the speed of light: electromagnetic sine waves

Every form of radiation, from sunlight to x-rays to AM radio, appears as electromagnetic sine waves that are comparable to sound waves except that

✔ An electric field is waving instead of air pressure, and

✔ The frequencies in Hz can range much higher than sound waves.

Just for another example of the tricks you can do with sine waves, I'll consider the case of AM radio. Here's the story: If you speak into a microphone, the sound waves are converted by the microphone into electromagnetic waves traveling down a wire. If you are singing a pure 500 Hz tone (you can't actually get it that pure), the microphone produces a pure 500 Hz electromagnetic wave in the wire.

The challenge of AM radio is to find a way to send this wave out on an antenna in a way that can be received miles away. It's done simply by multiplying the wave from the microphone with a much-higher-frequency *carrier wave*.

The basic idea is illustrated in Figure 14-13. For a station operating at 1,200 kiloHz (that's 1,200,000 Hz), the result is a wave that looks like a plain 1,200,000 Hz sine wave with an amplitude modified to display a 500 Hz ripple. The AM radio receiver can be tuned to subtract out the sine wave carrier, leaving an electrical 500 Hz signal to be passed to a speaker, where it's converted back to sound at 500 Hz.

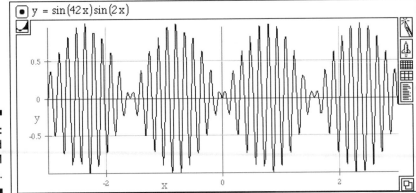

Figure 14-13: The grand plan of AM radio.

That's where AM radio gets its name; the AM stands for *Amplitude Modulation.* It's also the reason that AM radio isn't a first choice for music accuracy. Because there are stations spaced every 10,000 Hz or so along the dial, each station has to clip its music signals in frequency to prevent overlapping neighboring stations. There's plenty of signal space (called *bandwidth*) for talk, which doesn't require a big range of frequencies, but AM isn't going to do an outstanding job at violin concertos.

Anything you can pick up on any kind of antenna was originally assembled back at the station as a collection of sine waves added or multiplied together electronically, just like the sine waves in the little diagrams in this chapter. Half the tricks in modern electronics consist of ways to make the waves combine exactly without introducing distortion. And then, at the receiving end, you own equipment that can disassemble the waveforms back to a sound signal or a TV scan line or some other sort of message. The whole world of communications, including the satellite navigation signals previously mentioned, is based on nothing more than combining the humble circular functions of trigonometry.

Chapter 15

How Classroom Math Connects to Business Math

A good high school algebra course prepares you to do very well on the SAT test so that you can get into the college of your choice. It also helps you deal with problems involving trains headed toward each other at different speeds and their subsequent fates. It generally fails, however, in three key areas:

✔ Most of the work with formulas that you'll do in later life will be in setting up expressions that a computer will use. These expressions may be in a real programming language such as C or BASIC, but you may also have to deal with formulas in computer spreadsheets such as Lotus 1-2-3 or Microsoft Excel. Your algebra class probably offered no training in anything like this.

✔ You could figure out all the financial math in this book for yourself if someone had told you how to use the y^x key on the scientific calculator that you bought for algebra class. My guess is that nobody did. Most textbooks still treat calculators as an afterthought, with a few remarks relegated to an appendix.

✔ All sorts of fascinating developments in computer algebra have been roaring along in the 1980s and 1990s. Not only is this material interesting in its own right, but a computer-accelerated course in algebra could get to advanced topics (group theory and modern algebraic structures, for example) in a one-year course. It's highly unlikely that any of this gets mentioned in a regular high school algebra class.

And just to be as direct as possible about the shortcomings of traditional algebra course content, it's quite typical that even the better students aren't able to make basic financing decisions later. They can solve lots of goofy artificial problems about trains, but they can't tell when changes in interest rates mean that they should think about a mortgage refinance, and they believe amazing and fantastically contradictory probability results.

Business Meets Science on a Calculator

Although many of the things you learn in algebra are of importance mostly in physics and in subsequent math courses, some bits of algebra impact you every day.

Breaking even

In the endless quest to find some activity that makes money, you are constantly confronted with possibilities in which start-up costs and revenues are based on activities. These are just a few examples:

- ✔ You own a small graphics shop and are thinking of expanding your services. Should you buy a color copier? How many copies would you have to sell, and at what price, to make the copier pay for itself?

- ✔ You could buy a used truck and start a parcel delivery service. How much business would you have to do per week to pay for the truck in the first year?

- ✔ How many silk roses at $5 apiece do you have to sell as an annoying mime at a Renaissance festival vending booth to pay back the $120 that you spent on materials and the $30 vendor's fee?

You can solve these problems by using direct algebra or by graphing the solutions, which gives you a nice picture of the break-even points in these problems. Figure 15-1 shows the break-even point for the first example. A graphical solution makes a strong argument in a meeting, even if most of the people at the meeting can't follow simple algebra.

A huge fraction of all the computer spreadsheets humming away on the desktops of corporate America are figuring out break-even points for models that keep track of dozens of factors. Similar models are used to *optimize* results, rather than merely break even. When an oil company evaluates its options, for example, it has to keep track of the price of a wide range of refinery products, from gasoline to wax to plastic feedstocks, and adjust the refinery output to maximize its revenues.

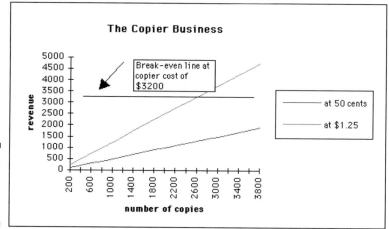

Figure 15-1:
The break-
even
problem.

Paying off

Back in Chapter 2, I outline some of the algebra involved in producing a formula for payments. Actually, the formula is a pretty complicated business, because interest is being charged while you are making payments and the amount of interest versus principal is changing with each payment. The simplest way to solve any problem involving time payments is to use a financial calculator, the second simplest is to use the functions in a computer spreadsheet, and the hardest way that actually still works is to use a scientific calculator with a y^x function key.

The payment formula for mortgages and ordinary consumer loans looks like this:

$$PMT = PV \times \frac{i}{1 - (1 + i)^{-n}}$$

For a bit of later-in-life algebra, you can try plugging in some numbers. I'm going to use the example of a car payment. I'll try a car that costs $12,500 and a payment plan that goes to 4 years at 9.1 percent interest. This example is almost realistic, at least in the 1990s.

The symbol *PV* in the formula stands for *principal value*, which is just the loan amount, so

$$PV = 12,500$$

n is the number of periods in the loan. It could have been 4 for 4 years, but interest on these consumer loans is always computed monthly, so

$$n = 4 \times 12 = 48$$

because there are 48 months in the 4 years.

i is the interest per period. That means that the number to plug into the formula for i is

$$i = {}^{0.091}/_{12}$$

$$= 0.00758333$$

That is, 9.1 percent is the same thing as the number 0.091, and you have to divide it by 12. Also, you have to keep a ridiculous number of decimal places in interest, since this interest number gets multiplied by itself many times and errors can start to accumulate.

Putting the pieces together, you get

$$(1 + i) = 1.00758333$$

Calculating $(1 + i)^{(-n)}$ is the exciting part. It's

$$(1 + 0.00758333)^{-48}$$

You enter the number 1.00758333 in the calculator, press the y^x key, enter −48 (which you do by entering 48 and then pressing the +/− key), and press the = key.

That complicated factor works out to

$$(1 + 0.00758333)^{-48} = 0.69584621$$

Plugging in the terms, you have

$$PMT = 12{,}500 \times (0.00758333)/(1 - 0.69584621)$$

$$= (94.791666)/(0.30415790)$$

$$= 311.65$$

Run through this kind of calculation a few times in an afternoon, and you'll be running down to K-Mart for a basic financial calculator that just has keys that say *i, n, PMT,* and *PV.* If you're clever and organize your work carefully, you can avoid entering these long numbers more than once, but if you're not that clever, it's really a pain.

Anyway, this exact formula is programmed into calculators and spreadsheets, and if you have lots of patience, you can derive it from simple considerations about interest compounding.

Geometry and Reality

Euclidean logical proof meets with little application in daily life, mainly because so few real-world axioms can be trusted absolutely. On the other hand, the actual physical world is built from the formulas of arc length, diameters, areas, and volumes that everyone learns in geometry class.

The gritty details

Suppose you have a space near the back door of your house and want to put in a concrete area that's 9 feet × 12 feet. This area will be a great assistance to fire safety when you use whole quarts of lighter fluid to ignite mere handfuls of charcoal briquettes for summer barbecues. Consider the geometry problems involved:

You need to construct a wooden frame for pouring the cement. For a nice, thick slab, you may want to use 2×8 boards. Of course, following the time-honored customs of building-material practice, a 2×8 is actually somewhat smaller than 2 inches × 8 inches in cross-section. You are making a frame that's 9 feet × 12 feet, so you need to buy the amount of lumber shown in Figure 15-2. That gives you the perimeter of the frame, 9 + 9 + 12 + 12 = 42 feet.

Now you need to figure out how many bags of cement to buy. The diagram in Figure 15-3 sets up the calculation. One of the first things you see (and when you think about the volumes involved, it makes sense) is that having a pickup truck to lug all that cement back from the hardware store would be a great convenience.

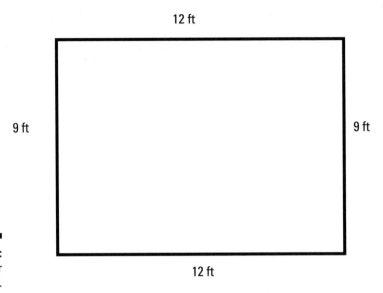

12 ft

9 ft   9 ft

12 ft

Perimeter = 12 ft + 12 ft + 9 ft + 9 ft = 42 ft

Figure 15-2:
A perimeter
question.

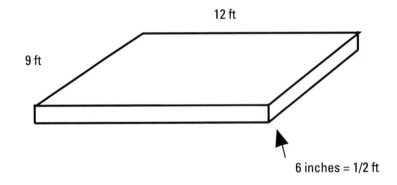

12 ft

9 ft

6 inches = 1/2 ft

Figure 15-3:
Cement
volume.

Volume of cement is 12 ft x 9 ft x 1/2 ft = 54 cubic feet

Now suppose you got really artsy and decided to try for the fancier shape
shown in Figure 15-4. Hey, people do stuff like this all the time. Again, you
would be calling on old geometry rules, and with a little bit of attention, you
would notice that you are adding the volume of a very shallow cylinder to that
of a very shallow rectangle.

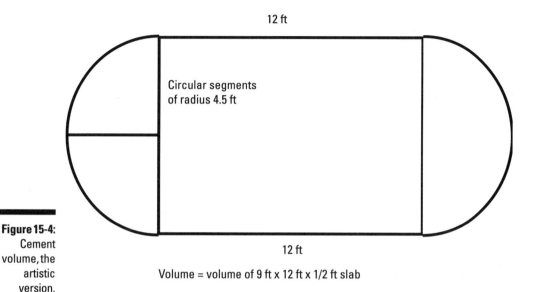

Figure 15-4:
Cement
volume, the
artistic
version.

Volume = volume of 9 ft x 12 ft x 1/2 ft slab

+ volume of cylinder of radius 4.5 ft x 1/2 ft height

A world of impossibilities

One of the most interesting features of geometry, although it may not have been discussed extensively back in school, is that outside the world of direct measurement, geometry generates problems of mind-boggling difficulty.

For example, once you try to cover an area with tiles that aren't rectangles or squares, you find thorny problems all over the place. Can you figure out how to cover a flat surface with the kind of tiles shown in Figure 15-5?

The simplest problems turn out to be very challenging. If you fill a cylinder with spheres, what fraction of the space in the cylinder will be occupied? If you fill a rectangular box with little cylinders, is that really the same problem as the sphere problem? Some of the problems in this class have been solved, while others, which you could easily state to a third-grade class, are still considered research problems. It's satisfying, in a way, that geometry generates so many interesting problems, often with no apparent business or practical applications. Euclid would be delighted.

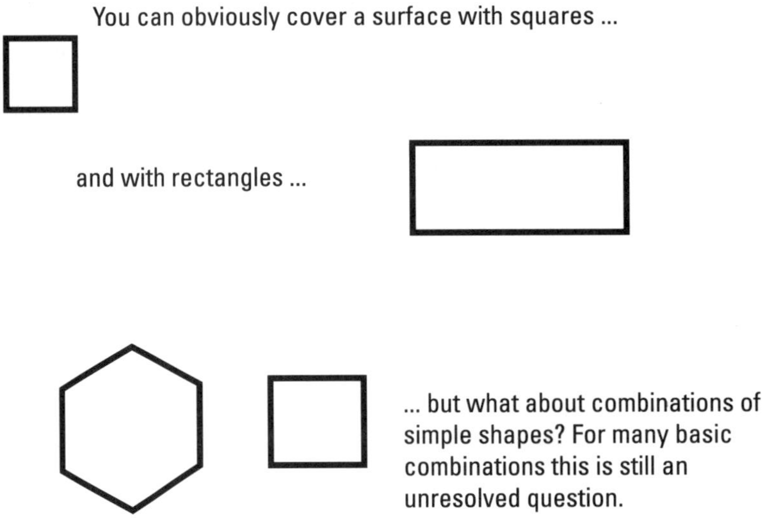

You can obviously cover a surface with squares ...

and with rectangles ...

Figure 15-5:
Exotic tiles
in geometry.

... but what about combinations of simple shapes? For many basic combinations this is still an unresolved question.

Thinking about Risk

Simple considerations of numbers and their relative sizes connect directly with daily life — in fact, that's one of the main themes of this book. Every day you are bombarded with information presented in terms of numbers, and if you think about the numbers a bit, you can decode the reality behind them. Often, this ability is crucial, as much of the information we get is distorted both by its presentation and by our own internal perceptions. When you make business decisions — for example, decisions about insurance — you make better decisions if you start with numbers that reflect reality.

For example, a perusal of the news on a daily basis might convince you that you are at immediate risk of death from murder, floods, riots, or accidents. You almost never have the opportunity to step back from the flow of news and ask yourself how any of the odds in these situations affect you. (See Chapter 19 for more insight into statistics in the news.)

Table 15-1 gives you a little homework table to photocopy. See if you can fill in the blanks over a period of weeks by paying attention to newspaper articles. Particularize the results to your own state or even to your own city if you can get the data. National statistics, for example, don't mean much — one set of numbers is appropriate for you if you live in Boise, and quite another set works for Houston.

Table 15-1	Your Own Risk Directory
See if you can fill in the blanks from newspaper and magazine reading. One blank has been filled in to give you some starting information.	

Event	*Risk*
Serious car accident	?
Commercial airplane accident	1 in a million per 10 hours
Cancer (lifetime)	?
Heart attack	?
Skydiving (per event)	?
Murder by stranger	?
Home accident (all kinds)	?
Natural disaster	?

Looking over the numbers, you may find that you have been worried about getting killed in a carjacking, whereas your real problem is that you smoke two packs a day. Practically everyone in California, for example, is at thousands of times more risk from drunk drivers than from earthquakes, despite rampant earthquake paranoia. Can you determine the difference in the chances of being shot in the best and in the worst neighborhood in your area? Should you relocate your business, or is your current location safe enough? Take it from me: The numbers are worth collecting. Even if you don't use it much, a perspective on numerical reality is a helpful thing to keep on hand.

Part IV
Off Duty

In this part...

I didn't provide you with all those terrifying credit card calculations just so you could get out of the time-payment trap and lose it all at blackjack. In my relentless concern for your well-being, I'm going to explain all the calculations in the background of daily life.

As always, my emphasis is on using the math to get a few basic principles right. Just to take the example of blackjack, I don't think that you should bother to memorize 15 more rules for special betting cases to reduce the dealer's margin from 1.5 percent to 1.38 percent. Similarly, I think a very simple rule that gets your restaurant tips right to the nearest dollar is better than a three-part mental-math trick that tells you to leave exactly $5.87 on a $39.13 check. Hey, live a little . . . leave 'em six bucks. You're here to do "good enough" math, which you will find not just delightfully easy but, in fact, plenty good enough.

Chapter 16
Tipping

*T*he United States is a unique national culture. It's not just apple pie, Mom, and Chevrolet — they have mothers and apple pie (of sorts) and General Motors cars that look just like Chevy Luminas in Germany. The really unique feature of the United States is its social organization. That means that a chapter on tipping (mostly in restaurants) is not just going to be a mental-arithmetic exercise about multiplying numbers by 15 percent in your head. In fact, at no point will I tell you to multiply anything by 15 percent. It's going to be much *easier* than that.

No, tipping brings up one of the most taboo subjects in America, the subject of social class. Most other countries have settled, as a custom, on a tip calculation that's built into your restaurant check. Other countries, with a more resolutely egalitarian approach, pay people a reasonable salary for working in restaurants and thus eliminate the need for the practice of tipping. In the land of the free and the home of the brave, however, tipping is up to you. But tipping is enmeshed in a detailed set of expected practices, so in real life your options are limited by social convention. Let me try to put this subject in context.

Picture This: A Restaurant Scenario

You're on vacation in Los Angeles. Emboldened by something you read in an airline magazine on the way, you decide to call Wolfgang Puck's famous restaurant, Spago, and ask for a reservation.

Ordinarily, your chances of getting a reservation would be less than outstanding. The presentation of restaurant protocol in the movie *L.A. Story,* in which the prospective customer has to submit to a financial examination, is not far off the mark, at least for the highest high-end places. But it's the New Hollywood. Sony invited everyone who counts to a party two days ago, and because Sony's studios have been losing money, the in-house caterers decided to recycle some sushi for the party. The entire celebrity guest list of Spago and its competitors has been brought to its knees in a most unseemly fashion, and restaurant reservations all over town have been canceled. In a once-in-a-lifetime miracle, unbeknownst to you, the staff at Spago is just delighted to be getting calls from Tom Smith of Zircon, Ohio.

Brad will be your waiter tonight, and he'd like to tell you a little bit about the specials. At the end of the recital, you remark that you could swear you recognized his voice from somewhere. So Brad says, "Yes, it's the story of my life. I have a brother who's a TV producer; he gets me voice-over jobs in commercials, and he makes me do a lot of bit-part work on the soaps. I have to do it because he's family and Mom gets upset when I turn him down. But I think my real calling is waiting tables, and I really resent having to interrupt it for acting."

Anything wrong with that scene? Think you'll ever hear that bit of dialogue? Even if you have never been an actor or a waiter, you can assess the probabilities because, although all men are created equal here in the U.S., we all know the occupational pecking order.

Six-year-old kids know that bank presidents "outrank" waitrers, that doctors outrank gas-station attendants — for that matter, most six-year-olds can tell you that Macy's outranks K-Mart. This puts a strange class-rank spin on restaurant dining that's not present in other standard commercial transactions.

The whole class-rank thing introduces some uncertainty into behavior in restaurants. For about 10 percent of the population, apparently, it represents an opportunity to bully people who are simply supposed to take it. Some restaurant customers feel a morbid concern about "doing the right thing," from table manners to tip amount.

The aim of this chapter is to help you relax. I suggest that you simply short-circuit all decision-making and follow the tipping rules that I present in the following sections. You won't have to worry about correctness, you will duck out of the weirdly artificial master-servant model of fine-dining service, and you will become a better person.

The Basic Tipping Rule

This rule applies to restaurant dinners or lunches in the range of $20 to $100. This range covers nearly 90 percent of the restaurant dinners for two in the U.S., according to the standard trade magazines on restaurants and institutional dining. By the way, read restaurant trade magazines only if you want to find out how much once-frozen chow is going down the hatch at some of this nation's finer restaurant chains. Watch out for the cannelloni at steak houses, if you catch my drift.

Without further ado, here's the rule:

The Rule of Two:

Look at the first digit of the total on the check.

Multiply by two.

That's the tip.

"Huh?" I hear you inquire. "What? No fumbling with complicated math?" Nope, it's that simple. Look at a few examples:

- The total comes to $45.61. The first number is 4. Four times two is eight. Leave eight bucks. That works out to be a 17.5 percent tip instead of the now standard 15 percent, but at this rate, it's only about a dollar more than an absolutely correct 15 percent.

- The total comes to $58.14. The first number is 5. That means that the tip is $10, which in practice is a 17 percent tip.

- The total on the check is $37.50. The first number is 3. The rule says that the tip is $6. That's a 16 percent tip, compared to an exact 15 percent tip of $5.63. What were you going to do — count out the pennies?

- The check says $28.10. The rule says that $2 \times 2 = 4$, and that's the tip. In this case, it's only 14.1 percent, but that's close enough for nongovernment work.

- You got appetizers, you got a decent bottle of wine, and the total came to $91.45. The rule say the tip is $9 \times 2 = $18. That's almost a 20 percent tip (19.7 percent), but face it: You're in a nice place, and the social fact is that 20 percent is beginning to be the custom in the pink-tablecloth, three-forks world.

The table in Figure 16-1 explains the whole story. Using the basic tipping rule, at least in the check zone of $20 to $100, your tips will range from a low of 13.3 percent (on the rare occasions when the bill is exactly $29.99) to a high of 20 percent, which occurs when the bill is an exact multiple of ten ($20, $30, and so forth).

Check amount	"Rule of Two" tip	Exact 15%	Exact 20%
$32.00	$6.00	$4.80	$6.40
$35.00	$6.00	$5.25	$7.00
$38.00	$6.00	$5.70	$7.60
$41.00	$8.00	$6.15	$8.20
$44.00	$8.00	$6.60	$8.80
$47.00	$8.00	$7.05	$9.40
$50.00	$10.00	$7.50	$10.00
$53.00	$10.00	$7.95	$10.60
$56.00	$10.00	$8.40	$11.20
$59.00	$10.00	$8.85	$11.80
$62.00	$12.00	$9.30	$12.40
$65.00	$12.00	$9.75	$13.00
$68.00	$12.00	$10.20	$13.60
$71.00	$14.00	$10.65	$14.20
$74.00	$14.00	$11.10	$14.80
$77.00	$14.00	$11.55	$15.40
$80.00	$16.00	$12.00	$16.00
$83.00	$16.00	$12.45	$16.60
$86.00	$16.00	$12.90	$17.20
$89.00	$16.00	$13.35	$17.80
$92.00	$18.00	$13.80	$18.40

Figure 16-1: A demonstration of the "first number times two" tipping rule.

Social justice, au jus

Here's a way to look at this simplified tipping procedure: You will be tipping at an average of about 16 percent. Unless you eat in good restaurants every day, this practice will add a trivial amount to your lifetime expenditures. At least in your final moments on this planet, you can reflect that you gave a little bit more money to some people who actually needed and appreciated it. If you follow the advice about managing your checkbook and your credit cards that I give earlier in this book, you will have more than enough money to overtip by a dollar or so occasionally. Think of the tipping rules and the other advice as a sort of package deal.

The IRS, in particular, now has procedures in place for estimating tip income for waitpersons. It makes no provision for times when waiters or waitresses get stiffed, nor does it make provision for the occasional senior citizen who has been tipping 25 cents since 1937 and is not about to change now, thank you very much. You will be helping to bring up the statistical average of tips received to the total that the IRS expects and upon which it bases withholding and audits. In other words, you will be giving a break to people who (usually) deserve it.

The Rule of Two: Some Follow-Up Points

Here are a few more considerations for fine-tuning the rule to your own circumstances.

Backing off a bit

If you pay much attention to arithmetic, you may notice that this rule has you tipping 20 percent when the check total is a multiple of $10. That is, when the check is $60, the rule says to leave $12, which is obviously 20 percent.

You may simply not feel like leaving 20 percent; it isn't really expected of you (not yet, anyway). So just back off the amount by a dollar or two. That's not much of an arithmetic problem, is it?

Now, in applying this rule, I ask you to keep in mind that in most cases the wait staff can't do much about a slow kitchen — if anything, they're being bothered more by kitchen backup than you are. As a general rule, if things have been ridiculously slow, you'll get better results from complaining to a manager than from stiffing a waiter on a tip. The manager may actually be able to do something for you, such as apologize and hand you a gift certificate.

The famous sales tax rule

In some states, where political cowardice and irresponsibility have led governments to rely ever more heavily on sales taxes for collecting funds (sales taxes are intrinsically the least fair form of taxation, taking the biggest total tax bite from the poorest people), the sales tax rates have reached numbers as high as 7.25 and 8 percent. In those cases, a simple rule for getting an approximately correct tip is to multiply the amount of sales tax on a dinner by two. This technique works in San Francisco, and it works in New York, but it requires a bit more multiplication than the Rule of Two. I only want you to multiply a single digit by two, and mah rule works in Texas, pardners. Y'all jes' think about *that* fer a minute.

Simple addition

One advantage of the basic tipping rule is that it gives you the nearest correct whole number of dollars. If you're leaving currency, you can just leave this number of dollars and whatever change ensues, and everyone in the whole place will think that you're unspeakably suave and worldly wise.

If you're putting the tip on a credit card, the rule here saves you some additional arithmetic, because the last two digits of a Rule of Two tip are zeroes. Consider the following two approaches.

The Rule of Two way

The check comes to $47.58. The first digit is a four, so the tip is $8. Your credit card arithmetic is

	$47.58
Gratuity (what a great word!)	$ 8.00
Total	$55.58

The official 15 percent way

You need to take 15 percent of $47.58. Doing the arithmetic in your head, you note that 10 percent of the total is $4.76. Five percent would be half that much, or $2.38. Adding these two numbers gives you a tip of

$$\$4.76 + \$2.38 = \$7.14$$

and a final check calculation of

	$47.58
+	$ 7.14
Total	$54.72

Your tip, calculated the Rule of Two way, represents a 16.8 percent tip. But you can do the calculation while you're talking, you can't get it wrong, you reduce your arithmetic load, and restaurant practice is drifting up a bit from 15 percent anyway. This technique lets you leave 15 percent plus change without knowing a mental-arithmetic trick for multiplying 17.5 percent by a four-digit number.

Fifteen percent is not a sacred number

The Rule of Two way has you tipping in a range that's correct for modern American restaurant practice. It is rarely exactly 15 percent. But remember that your waiter is not the IRS. When the Feds say that you're in an 18.5 percent bracket, they really frown on approximation. You don't want to give them 14.7 percent, nor is 19.3 percent a good substitution. In restaurants, you can expect more slack. After all, the customer is right about 70 percent of the time.

Tipping at the Ritz

Perhaps you have benefited immensely from the advice in the first part of this book (remember, one item of advice was to get a copy of *Personal Finance For Dummies* and do what Eric Tyson tells you). You are now rolling in money and are thinking of installing a swimming pool full of gold coins (this image has haunted millions of people familiar with the name Scrooge McDuck from childhood readings of Disney comics). You've taken to dining frequently at the Ritz, by which I mean the *real* Ritz-Carlton in Boston, not one of these West Coast Ritzes where people sit around in the lobby in tennis shorts.

You may be entirely familiar with the concept of a small salad of mixed baby field greens for $14 and rarely drink a wine that costs less than $50 a bottle in restaurants. This, in turn, implies that your typical dinner for two runs higher than $100.

Or perhaps you're supposed to pick up the tab for a group by using the company credit card. Here's the rule for restaurant tabs over $100:

> **The Ritz Rule:**
>
> **Look at the first digit of the check.**
>
> **That's how many $20 bills you leave.**

If you have trouble with that rule, the company needs to give the credit card to someone else, and it's a real mystery how you could be out there lifestyling on your own. Figure 16-2 confirms this rule.

Figure 16-2: High-end tipping and the Ritz Rule.

Big-Ticket Tipping			
Check amount	**"Leave the $20s" tip**	**Exact 15%**	**Exact 20%**
$200.00	$40.00	$30.00	$40.00
$225.00	$40.00	$33.75	$45.00
$250.00	$40.00	$37.50	$50.00
$275.00	$40.00	$41.25	$55.00
$300.00	$60.00	$45.00	$60.00
$325.00	$60.00	$48.75	$65.00
$350.00	$60.00	$52.50	$70.00
$375.00	$60.00	$56.25	$75.00
$400.00	$80.00	$60.00	$80.00
$425.00	$80.00	$63.75	$85.00
$450.00	$80.00	$67.50	$90.00
$475.00	$80.00	$71.25	$95.00
$500.00	$100.00	$75.00	$100.00
$525.00	$100.00	$78.75	$105.00
$550.00	$100.00	$82.50	$110.00
$575.00	$100.00	$86.25	$115.00
$600.00	$120.00	$90.00	$120.00

Just consider these examples:

- Simple little lunch, followed by a couple of 1914 Hine cognacs, which tend to run $45 apiece. Hey, they're running out of the stuff, a lot having gone down the hatch at the 1918 Armistice.

  Total on check = $141.50

  First digit says 1, so you leave $20.

  You just tipped 14 percent, but that's okay considering the relatively light service demands in delivering two cognacs.

- Business dinner for four at the "good" Japanese restaurant a few blocks from the convention center.

  Total on check = $317.40

  First digit says 3, so you leave three twenties for a tip total of $60.

  You just tipped 18.9 percent, a perfectly respectable amount under the circumstances.

The only place this rule seriously undertips is in the region of $160 to $200. As you get close to $200, you should throw on an extra $10 bill, even if you're still below the $200 mark. It's not a problem up there in the $200-and-beyond stratosphere, but at $195, it's been a long time since the tip amount was updated (you've been leaving $20 since the tab was $105).

Party of Four Tipping

There's a variation on this trick for splitting the check in a party of four, including the tip. It works quite well for amounts from $40 to $100, so it's pretty good for lunch with coworkers. If you know a good restaurant where the final total for a party of four comes to less than $40, send it to me in care of IDG Books Worldwide, Inc., and we'll print a list in the next edition of this book.

Party of Four basic rule:

Look at the first digit in the check.

Multiply by three.

Party of Four rule, part 2:

Round up the check amount to the nearest ten first if the check is halfway there.

Here are two examples:

- The check is for $48.79. This amount (look at the second digit) is more than halfway up the ladder to the next ten, so round it up to $50. Everyone thus pays $3 \times 5 = \$15$. This method produces a generous 22 percent tip, but it preserves tranquillity in your group. If you object in principle to giving a bit more than 20 percent on occasion (it happens here because of the round-up), you (as the group leader) can take everyone's cut down by a dollar. That makes every payment $14, and the tip is 14.7 percent.

- The bill comes to $53.11. The first digit is 5, so you multiply by 3, and everyone pays $15. Leaving 4×15 ($60, but you don't even have to calculate that part) means a tip of $6.89, which is a 13 percent tip. On the occasions when you don't round up, you can have everyone throw in another buck if you're in a good mood. That brings the tip up to 20 percent.

In general, you can fine-tune the tip by adding or subtracting a dollar. With a check total like $72.23, you're effectively rounding down when you multiply seven times three, so figure that you're going to supplement the tip amount with a dollar (unless you got coffee poured on your lap). In the other direction, if you have to round *up,* you can pull a dollar per person back off the pile.

If you're in a party of four and the total is over $100, just take the first two digits, multiply by three, and forget about rounding and dollars and the other fussing:

- On a check for $185.60, take the 18 and multiply by 3. Everyone pays $54.

- On a check for $151.17, take the 15 and multiply by 3. Everyone in the group pays $45.

How about Six?

There's a similar rule for parties of six, except that you can't always use it. Over the years, restaurants have had problems with severe undertipping, so in many cases (especially upscale places in big cities), you may see the following message in tiny letters at the bottom of the menu:

A gratuity of 17 percent will be charged for parties of six or more.

In that case, you just have to divide by six. And I'm sorry to report that there really isn't a trick for dividing by six that isn't as much trouble as doing the original division.

If a tip isn't automatically added, there's a rule just like the Party of Four rule. The magic factor here is two, not three. Because parties of six result in larger restaurant tabs, typically over $80, you can forget about making little $1 adjustments. Here are a couple of examples:

- The check for the six of you is $92.23. The first digit is 9, so everyone kicks in $2 \times 9 = $18. The total payment is $6 \times 18 = $108, meaning a tip of 17 percent. Okay, it's really 17.1 percent, but it's good enough.

- The check says $150.80. You say 2×15. That means $30 apiece for everyone in your group. You are thus leaving $180, which amounts to a 20 percent tip. It's 20 percent because the check has stopped on a multiple of ten, so there's no tip-shaving from rounding down a check.

What's the basis of these tricks?

If you don't mind a bit of algebra, I can show you a simple basis for all these little gimmicks. What's going on is that I recommend a really easy calculation (multiplying by two) on a rounded-off number. It just works out that the rounding-off almost always leaves a tip in the range 15 to 20 percent.

The Party of Four and Party of Six tricks are really the same thing. Suppose you're in the party of four and you get the check for an amount I'll call *Check*.

- A 20 percent tip takes the total from Check to (Check + 20 percent) $\times$ Check, which is (1.20 $\times$ Check).

- You have six people present, so the individual share is $^1/_6 \times 1.20 \times$ Check.

- That multiplies out to $0.2 \times$ Check. So taking the tens digit in the Check amount and multiplying by two is exactly the same as using the factor $0.2 = {}^2/_{10}$.

So, again, what you're really doing in all these cases is making a first guess at a 20 percent tip and counting on a rounding procedure to shade this tip amount downward. You can see for yourself that, for dividing a check plus tip among three people, the magic number is 4. You and your pals get a check for $42.14 — all of you kick in $4 \times 4 = $16. If the bill comes to $48.30 instead, you round up to $50 and pay $5 \times 4 = $20 each.

Little Tips: Coffee Shops and Cabs

There are two ways to figure these kinds of tips, and you can take your pick.

- ✔ You can round off the amount of the check or cab fare to some number that's convenient for you. Maybe you remember that 2×12 is 24 but for some reason have trouble with 2×13. Then multiply this dollar amount that you remember by two. If you rounded up to make things convenient, you then can round the tip down to the nearest dollar.

- ✔ You can multiply the dollar amount by two and tip at the closest dollar. For example, if the fare is $6.70, multiply 6.70×2 to get $13.40. Rounding off, you get $13, so tip $1. Especially in cabs, do everyone a favor and tip in whole dollars. In this case, you may want to give the driver $8 ($6.70 + $1 = $7.70, rounding up to get $8).

Last Words about Tipping

There are lots of little ways to save money in life and lots of little ways to spend extra. As a practice in math and money, I want you to overhaul your consumer debt and avoid bounced-check charges. You can save thousands of dollars per year by getting credit card balances out of the 20 percent interest zone down to 9.5 percent. You can save hundreds by keeping your checking account in good shape. And this will give you the opportunity to overtip, very slightly, by about $60 to $100 dollars in total over the course of a year. Winning the bigger money battles with institutions so that you can afford to be a bit more gracious to the people you meet in everyday life just seems to be a better way to do things.

The 5th Wave
By Rich Tennant
WAYNE GETS A SUDDEN LESSON IN HOW TO CALCULATE A CORRECT TIP AT A RESTAURANT
©RICHTENNANT
That should be enough.

Chapter 17

Gambling

• •

In This Chapter

▶ Flipping a coin: the simplest kind of bet

▶ Turning a dollar into a quarter by buying scratcher cards

▶ Playing roulette

▶ Avoiding disaster in big bets with the Wilcox principle

▶ Understanding slot machines, keno, and lotteries

▶ Playing it smart: basic rules for blackjack and poker

• •

I'm now going to give you some advice about gambling — and life, for that matter. Let me approach the subject in an uncharacteristically roundabout way.

The Basics

Let's play a little game. You and I take turns flipping quarters. You go first. If your quarter turns up tails, you give it to me. If it comes up heads, I give you a quarter. Then it's my turn to do the same. We continue until one of us runs out of quarters.

Does the payoff in this game sound about right to you? It probably does, because nearly everyone can understand that because a coin has two sides, there are two possibilities. The payoff if you call the bet right, then, is two to one (you end up with two quarters if you bet one of them).

I would confidently assert that you could tell similarly fair bets from less-fair bets, except the probabilities in nearly all "interesting" games get muddled by a much larger assortment of possibilities.

The key concept here is *expectation value*. The expectation value roughly is the payoff on a bet times your chances of winning it. In the case of the quarters game, you pay 25 cents to play a game with this expectation value:

expectation value = payoff × probability

$$= \$0.50 \times \sfrac{1}{2} = \$0.25$$

You pay a quarter to play this game, and the most likely outcome is that you'll get a quarter back. Neither player has an advantage or an "edge" in this game. It's just about the working definition of fair, as far as gambling goes. For reference, the odds and the payoffs in the "quarters game" are summarized in Figure 17-1.

Figure 17-1:
The quarters game: the rules, the odds, the payoffs.

The Quarters Game

			probability
You flip	tails	result = pay me a quarter	0.5
	head	result = I pay you a quarter	0.5

			probability
I flip	tails	result = I pay a quarter	0.5
	head	result = you pay me a quarter	0.5

Suppose we both start with 20 quarters and do a series of 20 flips. The graph in Figure 17-2 shows the set of possible outcomes. What typically happens is that we both break even, plus or minus a few quarters. In rare cases, one or the other of us may get ahead by a few dollars. The odds that you will win all 20 quarters from me are almost exactly one in a million.

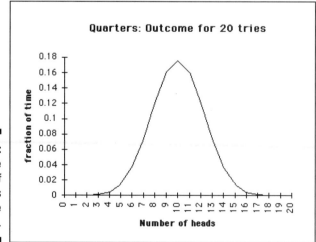

Figure 17-2:
Possible outcomes of the quarters game: the long run.

Scratching Your Way to Riches?

The game of matching quarters has an expectation value of 25 cents for every quarter you play. It's a fair game. Since expectation value is a good way to evaluate a gambling proposition, you can use it to evaluate the little games your state would like to play with you.

California and many other states sponsor little point-of-sale gambling games. A popular form is the Instant Winner "scratcher," in which you pay a dollar for a little card, scratch off some silver areas looking for matching pictures or some other payoff indication, and get paid on the spot. Figure 17-3 shows you a sample of one of these cards — they're sold in vending machines in stores all over the state.

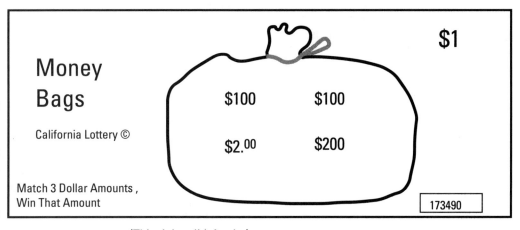

$1

Money
Bags

California Lottery ©

$100 $100

$2.⁰⁰ $200

Match 3 Dollar Amounts ,
Win That Amount

173490

Figure 17-3:
California
Instant
Winners? A
sample
scratcher.

(This ticket didn't win.)

On the back of the little scratcher card is an explanation of your chances. It couldn't be more explicit. The fine print states in fairly plain language that your expectation value on a one-dollar bet is *25 cents*. That's the California payoff, and it's roughly what you get in most other states in this kind of game.

In other words, when you buy a scratcher card, you turn a dollar into a quarter. Your expectation value — your most likely long-run payoff — on a one-dollar bet is just a quarter. *Am I making myself clear??!!*

One of my personal heroes in statistics is a gentleman who made newspaper headlines about eight years ago. He took a $10,000 retirement settlement and, with great hopes, bought $10,000 worth of Instant Winner scratchers. To his bitter disappointment, he found after scratching all those little cards that he was standing in a pile of debris worth $2,648.

Ancient and modern history

The state-sponsored gambling system in California is not taking in nearly as much money as its proponents expected. I think that's because you can somehow tell that you're not winning on scratchers even if you can't count on your fingers. You buy one, you lose. You buy one, you lose. Sooner or later, you get the picture, even if you have a bad memory.

People just *lose interest* if they don't get hits frequently enough, and the gambling games that have gone on for centuries have incorporated this information. There are reasonably good records of lottery-type games in the big cities of the Roman Empire. The evidence is that the people running the games traditionally kept about half the money and paid out the other half in winnings.

In the slums of Rio, playing the "animal game" (an adaptation for a nonliterate population) gives about the same expectation value. Inner-city numbers rackets traditionally pay out about 40 percent of the take.

The people who operate TV game shows also understand this idea perfectly. The show *Jeopardy,* I can state from my own fabulous personal experience, does mountains of research and testing to assure that the contestants get the right answer about $2/3$ of the time, and the average member of the home audience gets the answer perhaps 40 percent of the time. Takes lots of balancing Flintstones versus Shakespeare questions, but it's worth it.

Because this is a game of chance, it would have been remarkable if this person had won *exactly* $2,500. In fact, although he was hideously disappointed, he had excellent chances of getting *less* than $2,500. As it happened, he proved by using ten grand of his own money that, on the average, each one-dollar scratcher is worth a quarter. Presumably, if this civic-minded gentleman had wanted to lay it all on the line in the interest of applied probability, he could have rounded up $100,000 from his friends and relatives. The return would almost certainly have been an even more exact approximation to a 1-in-4 payoff, or $25,000 plus or minus change.

The fellow in question complained that he thought he would have at least *some* big winners in all those cards. Hmm. It says right there on the back of the cards that the state, in its infinite wisdom, has taken that possibility into account. You ain't gonna win. Not in this game, anyway. You may even get some cards that are $500 winners, but these are sufficiently rare that, when the smoke (and the little piles of silver scratch-off dust) clears, you're not going to come out ahead. You are guaranteed to lose, and each card tells you so in black and white. It's one of the few promises that your state intends to keep. It was just kidding about the tax cut.

Reduced Payoffs on Longshots: How Casinos Win

Now consider the case of keno (a game that resembles bingo) in the casinos of Nevada. Certain keno combinations have odds of 2 to 1 and 5 to 1, and the casinos pay $2 and $5 on a $1 card in these respective cases. Fair enough. Other possible combinations have odds of 5,000 to 1 and 10,000 to 1. And on these bets, the casinos typically pay a lot less than the odds would suggest. In some cases, a casino might pay $2,500 on the 5,000-to-1 case and $4,000 on the 10,000-to-1 case.

It's as simple as this: As long as the casino ducks the big hits, it can pay out very fairly on the little hits. Over the long run, it can't lose. Because a few thousand dollars seems like a big deal to most people, the casino seldom encounters people with a handbook on probabilities complaining about being a big (if not big enough) winner.

The same principle applies to slot machines, although the casinos pretty much assume that slots players don't even deserve the courtesy of an explanation. Slot machines generally pay back quite a bit of the money they take in, partly because they must be *seen* to be paying off or else the patrons won't step up and start pulling. What slot machines *don't* do is pay out many big winners — the rate of payoffs at a thousand dollars and up is about a tenth of what it should be in a "fair" game.

Roulette: Another Kind of Game

A flipped coin is essentially a little random-number generator. You can look at a coin as generating the numbers zero (you lose) and one (you win). The scratcher card is really another kind of random-number generator. It generates the numbers one (you win) and two, three, and four (you lose). Of course, since no one is that fascinated by numbers, scratchers have cool pictures and so forth to make losing money interesting.

A roulette wheel is a fancier example of a random-number generator. The croupier spins the roulette wheel, and the little ball eventually bounces into one of the numbered slots. And since roulette is a suave sort of activity, at least at some places, you can feel yourself removed from the wretched, guaranteed-loser world of scratcher game cards in convenience stores.

Welcome, Mr. Bond

Here's how roulette works in the casino at Baden-Baden in Germany. You trade your Deutsche marks for little gold chips. The roulette wheel has numbers 1 through 36, alternately red and black. There's also a number 0 in green (see Figure 17-4). If you bet one gold chip on red and the ball lands on a red number, you get two gold chips. That's because the wheel is half red and half black (except for the zero), so it's an even money bet. If you pick a particular number, you get a payoff of 36 to 1 because there are 36 numbers (again, except for the 0).

Figure 17-4:
The European roulette wheel layout.

1	2	3
4	5	6
7	8	9
10	11	12
13	14	15
16	17	18
19	20	21
22	23	24
25	26	27
28	29	30
31	32	33
34	35	36
	0	

The casino makes its real money when the ball lands on 0. When it lands on 0, the casino collects all bets. Well, actually, it also collects about $8 for a plain soda water and nearly $20 for a small glass of wine. But collecting when the ball lands on 0 means that the house picks up all the money. That means that one time out of 37, on the average, or

$$\text{"house collects" number/all numbers} = 1/(1+36) = \tfrac{1}{37} = 0.027$$

of the time (nearly 3 percent of the time), the house collects all bets. In a night when the equivalent of, say, $10 million is bet, the house collects $270,000. The rest of the action just consists of redistributing gold chips from one customer to another.

The casino is content with 3 percent of the turnover. Your expectation value per dollar bet (they don't let you bet dollars — I'm just putting it this way so that you can compare it to other bets) on a single number is

$$\text{expectation value} = \text{payoff} \times \text{probability}$$

$$= \$36 \times \tfrac{1}{37} = \$0.973$$

You can play this game for a long time if you have a big stack of gold chips because the built-in drain-down rate is so modest. This game, despite its glamorous trappings, is very nearly as fair as the quarters game mentioned at the beginning of this chapter.

Welcome, Bobby Ray

Here in America, people do everything more dramatically than those subdued Europeans do. When you look at the American roulette wheel shown in Figure 17-5, one of the first things you may notice is that the wheel has not only a zero, but also a double zero. The casino has thus, by this shockingly simple artifice, very nearly doubled its take. On the average, the casino now collects

"house collects" numbers/all numbers = $^2/_{(2+36)}$ = $^2/_{38}$ = 0.0526

or a little more than 5 percent of all the bets in an evening.

1	2	3
4	5	6
7	8	9
10	11	12
13	14	15
16	17	18
19	20	21
22	23	24
25	26	27
28	29	30
31	32	33
34	35	36
0		00

Figure 17-5:
The
American
roulette
wheel
layout.

But many casinos in Nevada throw in an additional wrinkle. Often, the roulette setup features a little sign explaining the payoffs. "Why bother to explain the payoffs?" you may reasonably ask. "It's all self-explanatory." The sign contains valuable bits of information that you ought to pay careful attention to, however. It explains that you get a payoff of 30 chips (not 36) if you pick the right number. All sorts of other bets (betting on odd numbers, betting on even numbers, betting on numbers 1 through 18 as a group) in some cases also get clipped a bit, as you can see in the payoff table of Figure 17-6.

1	2	3	— This 12-to-1 bet sometimes pays 9-to-1.
4	5	6	
7	8	9	
10	11	12	
13	14	15	
16	17	18	
19	20	21	
22	23	24	
25	26	27	
28	29	30	
31	32	33	— This 36-to-1 bet sometimes pays 30-to-1.
34	35	36	
0		00	

Figure 17-6:
Roulette
payoffs
don't always
match the
odds.

What's your expectation value on a $1 bet on an individual number? According to the table, you might get a $30 payoff for picking the right single number. That means that your expectation value is

$$\text{expectation value} = \text{payoff} \times \text{probability}$$

$$= \$30 \times \left(\frac{1}{2+36}\right) = \$0.789$$

This little change in the rules greatly improves your chances of getting cleaned out in a single evening. Under the European rules, each pass at the wheel gives you an estimated net loss of 3 percent of your bet. In the American system, you have an estimated net loss of 21 percent of your bet (that's $1 — $0.79, rounding off). After ten American bets on single numbers, you should have about $1/10$ of your money left.

As a practical consideration, if you want to play roulette under these circumstances, just bet on red or black. On a 2-to-1 bet, you're only up against the ball landing on a green space, so you lose money much more slowly. By trimming the bigger potential payoffs, the casino severely reduces your chances of breaking even with occasional longshots.

It's in the Cards

Card games offer several opportunities for fair or nearly fair gambling — if played properly, your odds are better than they are in roulette. "Nearly fair" in this sense means that the bias against you on each turn is small.

In the game of matching quarters at the beginning of the chapter, you can see that the odds of winning on any toss are exactly 50/50. After a certain number of tosses, you may be either winning or losing, drifting up and down in money. In the long run, your average expectation from this game is zero, although when you stop at any point you can be up or down.

In "ideal" gambling, your results average to zero in the long run. In real-life ordinary gambling, in games like roulette, there's a slight drift on each bet in the direction of loss (the house "take"). In standard American roulette, you have 2 chances in 38 of hitting one of the green spaces. $^2/_{38}$ is about 5.3 percent, so you can expect to lose your starting stake after about 20 rounds just by betting on black or red. You might actually win in a single real-life set of events, but your expectation value, the most likely outcome, is to have about 5 percent less money per round. The game has a built-in drift in the direction of loss. It's not as bad a drift as state gambling, in which winning is like swimming up a water-fall, but it's a drift nonetheless.

Slot machines are the same sort of proposition. The "good" ones have about a 15 percent take, so you can start with a stock of 100 quarters and play for quite some time before the machine gobbles up all your money. You can expect to play something like 700 rounds on a good machine before you have worked through 100 quarters (assuming that you keep plowing your winnings back into the slot). The problem, of course, is that once you get rolling, you can run a turn in five or six seconds, so you're yanking the handle for about an hour as you dribble into oblivion. The two situations, a fair game with no bias and a real-world game with a built-in edge for the casino, are compared in Figure 17-7.

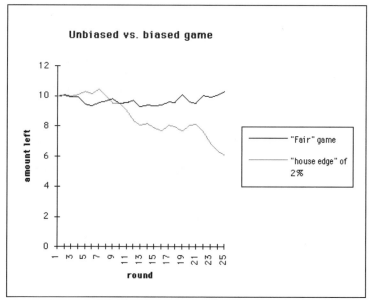

Figure 17-7:
Fair versus
biased
games.

The special case of blackjack

If you don't already know the rules of blackjack, I'm afraid you're not going to find them here. My advice is that you should learn the basics of such games in dimly-lit casinos with free drinks, not from math books. But I do have something to say about the odds.

Years ago, blackjack got a reputation as a game in which an astute player could actually get an advantage — a *slight* advantage — by carefully noting all the cards that were dealt. In the simplest instance of card counting, picture yourself keeping track of all the cards dealt from a single deck. If you get to the end of the deck, for example, and you have been keeping such careful track that you know that the last card to be dealt is going to be a ten, then you can predict exactly how the hand will turn out. Accordingly, if the ten is going to be dealt to you, and you have 11 points showing, you may be inclined to bet a bundle.

In a real blackjack game in a casino, of course, the cards are seldom dealt from a single deck. And if you make a giant bet with a grin on your face because you have succeeded in keeping count, you will likely find yourself, in just a few minutes, in the presence of several individuals chosen for their personal unpleasantness. These people will tell you to leave the casino. Casinos do not propose to lose large sums of money, and they don't care whether they live up to *your* idea of what's fair. After they throw you out for being too clever, there are still plenty of people who are willing to take your chair at the table and bet like idiots.

The simplest blackjack rules

Different types of computer analysis of blackjack have produced a fairly simple set of rules. There are only a few worth remembering, although some books on the subject list 30 to 40 rules and exceptions. If you're willing to memorize that much stuff, there are better uses for your intellect. For now, just try these:

- ✔ If the dealer shows a seven, eight, nine, ten, or ace, and you have a hard total of 17, then you should stand. (A *hard total* means the actual points on the cards; a *soft total* counts an ace as 11.)
- ✔ If you have a soft total of 18, stand no matter what the dealer shows.

Under the rules at most Western casinos, just following these rules puts you in a game that gives you a disadvantage of just a little more than 2 percent per round. That's pretty good by the standards of most of the gambling opportunities I've been discussing. Although there's a long-run bias against you, it's a pretty modest bias.

An argument for keeping it simple

Now here's the key argument. If you learn about 15 more rules for special cases, covering "doubling down" and "insurance" bets, and furthermore are willing to keep a count of tens and non-ten cards played, you can shave the house advantage to a bit less than 1 percent. The problem is, all this extra expertise hasn't bought you anything significant. You used to have a 52 percent chance of ending up with a net loss, and now you have a 51 percent chance of ending up with a net loss. This is still not a brilliant way to get rich.

Back when a fellow named Edward O. Thorp brought out his pioneering computer-based blackjack system in the 1960s (summarized in a famous book called *Beat The Dealer*), there were a few years in which diligent students of the game actually had an advantage of a few percent against the casinos. You still had close to a 50/50 chance of losing anyway, but you could put a slight bias in your favor, so you had long-run prospects of winning at blackjack.

Then the casinos changed the rules in response to the system, and they continue to change rules whenever they feel that it's necessary. Oh, and most large casinos have a few dealers around who are experts at cheating with cards. As a fun "thought experiment," imagine what would happen to you in Las Vegas if you accused your blackjack dealer of cheating (even if it's true). You don't need much math to assess your chances in this case.

The very special case of poker

Technically, poker is not gambling. According to a number of judicial decisions, five-card draw poker is a game of skill, not chance, and is therefore allowable under a different set of legal conditions from slot machines and so forth. If you can find yourself some poker-playing pals with bad habits (humming softly on straights and flushes, frowning at pairs of threes), you can do pretty well with minimal reliance on mathematics.

Using a computer, it's possible to generate a table with the odds on every poker hand. The catch is, you're not likely to remember the table in real-world circumstances. Sitting there around the kitchen table at a friend's house, with pound quantities of Doritos and six-packs of Bud disappearing every hour, you need something a little simpler. (I can *almost* picture this situation, except that here in Sonoma County we have little snacks made from Peruvian purple potatoes, drink champagne cocktails with cassis at poker, and are required by county law to wear burgundy velvet smoking jackets and monogrammed brocade slippers so that we can be identified as poker players.)

Thus, a great simplification is in order. Here it is, specifically for five-card draw, the customary "real" poker game.

Four-person poker

In a four-person game, you need to have a hand with at least a *pair of tens* to have a *50 percent* chance of winning.

If you really like to have things nailed down, you should have at least *two pair, kings high* to have a *90 percent* chance of winning.

If that's too much information, just remember that unless you're given some other bit of information — your opponents are weeping, for example, or giggling and putting poker chips over their eyes — you need a pair of tens to be in the running in a four-person game.

Six-person poker

In a six-person game, you need to have a hand with at least a *pair of queens* to have a *50 percent* chance of winning.

Once again, if you like nearly sure things, you need at least *two pair, aces high* for a *90 percent* chance of winning.

Actually, the exact number is closer to 86 percent certainty. If you're really paranoid, don't bet the farm in a six-person game without three of a kind.

The reason I'd rather give you these greatly simplified rules is that an approximate feeling for the odds, coupled with close inspection of the people playing against you, works more to your advantage than memorizing a math table. Some people bluff all the time, some people almost never do, and some begin tapping their feet slightly when anything exciting happens. On anything higher than a pair of sixes, I tend quite to forget myself and begin betting in a thick Lowland Scots accent, a sort of mock-Glaswegian. I don't recommend this.

As a last bit of statistical math advice, it happens that poker results on a given set of people are fairly reproducible. If you lose often in games with the same set of people, take up a different card game or else find different people.

Give Yourself a Break: The Wilcox Principle

This business about clipping the big payoffs is a clue to a larger lesson. I'll give you a roundabout background story, and I promise you a very generous expectation value if you follow it to the end!

There's a subtle board game called Go, played mostly in China and Japan (it's considered the Japanese national game). A very smart American computer scientist named Bruce Wilcox was trying to write a computer version of Go — like the computer programs that play chess — and came up with a principle that extends far beyond games.

The problem is that Go is lots trickier than chess, as Figure 17-8 demonstrates — there are 361 positions on the board, among other things. In chess programs, it's possible to have the computer evaluate most of the likely moves from a given situation on the board. The best chess-playing computer uses exactly this sort of brute force approach, playing nine or ten moves ahead. Go has many more possible moves than chess, and evaluating the results of moves ten turns ahead is out of the question even on the fastest computers you can imagine. So how do you generate reasonable Go moves?

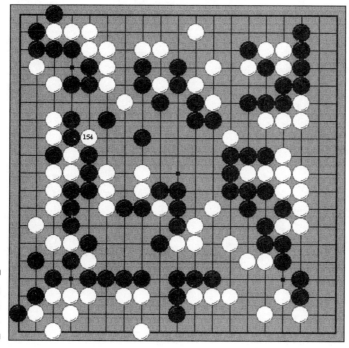

Figure 17-8:
A Go game
in progress.

Dr. Wilcox found that if he generated a fair number of random moves, played them out a few moves ahead, and eliminated the absolute disasters, the program would play near the level of the lowest professional rank (that's an amazing accomplishment). In other words, it didn't need to find the One Best Move at any given point — it just had to avoid the moves that worked out to be really stupid.

This discovery is of great significance. If you think about it for a while, you see that it generalizes to all sorts of situations, not just games. All the people who collect your money if you go gambling already understand it implicitly — they have all sorts of rules that prevent you from winning too much. In daily life, you are basically stuck playing an elaborate game with rules that change a bit each day and with expectation values for different moves that have built-in uncertainty. And there's only one way out of the game.

The Wilcox Principle

Simply put, the statistical result of Bruce Wilcox's research on Go is that you can do quite well just avoiding disasters. Perhaps I should illustrate this with specific applications.

- **High school:** Don't take a ride home from a party with a friend who is drunk. It's even 50 times worse if you take that ride in the back of a pickup truck. At this point, given the current relaxation in academic standards, you can just about guarantee that you will graduate from high school if you follow only this one rule.

- **Motorcycles:** Motorcycles are actually safer than people usually think. On a wet night, however, they're more dangerous than most bikers think ("Ya know, I've made that turn a hundred times, but all of a sudden the back end just slid out, and…"). The safety numbers change radically. You're not only invisible, but you also have almost no traction. Stick to cars or stay home if it rains.

- **Credit cards:** Never tell your credit card company to take a hike. Even if you get behind for a while and can't make minimum monthly payments, if you call up every month, announce any payment at all, and come up with a plan to cover at least the interest, you won't fall into oblivion. If you don't respond to letters, or worse yet tell them to whistle for it, you'll disappear off the credit radar for years, creating lots of nasty problems.

- **Life:** Don't marry someone who gets into violent arguments with you when you're dating. People collect statistics on this sort of thing, and the numbers say that you just shouldn't do it. After you sign the paperwork, the gloves *really* come off.

What emerges from computations and statistics on real life, as well as Go, is that you don't have to worry much about making exactly the right move. Just don't stick your head in a blender to see what it feels like. Finish school somewhere, show up on time, smile and nod a lot, avoid jerks when you can, and everything is likely to work out fine. I'm not just guessing here, and I'm not Joyce Brothers. I'm just reporting what the *math* says!

But please remember that, if you can absorb only one lesson from this chapter, that lesson should be what I will call the Wilcox Principle. It's this: *Dodge the really big losses, and you'll be okay.*

The Lottery

The people who run state lotteries understand the Wilcox Principle, even if they've never heard of Bruce Wilcox. In a lottery, the one big rule is this:

Take in lots more money than you pay out.

The main way lotteries adhere to this principle is that, even when the prizes appear huge, the payouts are pretty meager for the odds.

In a recent California lottery, the payoff was $17 million. If you followed the history of the buildup to this payoff, however, you know that the total amount of all lottery tickets sold was nearly $160 million (that's the number of combinations selected at $1 per pick). Assuming a standard bet of $1, this situation shows that, although the payoff was $17 million, the odds against you were 160 million to 1. To calibrate this probability, consider only that, as a healthy, nonsmoking 30-year-old, your odds of dying this year are approximately 1 in 1,000. Backing up a bit and rounding off as is my usual habit, I can tell you this:

You were a hundred thousand times more likely to die this year than to win this lottery.

Now, does *that* sound like a sporting proposition to you?

Personally, I know lots of people who buy a lottery ticket once a week and regard it as an entertainment expense. As habits go, buying a ticket now and then is probably better than racing motorcycles on weekends. If you realize that the odds are against you and don't count on the money showing up, a few tickets are a pretty harmless expenditure.

Chapter 18

Sports

Some of the material in this chapter may violate your existing preconceptions about sports activities. You may find my conclusions unbelievable. Just so you're warned.

If you want to disregard this bit of sports math, I'm going to help you by making a confession. I was born in Chicago and spent my formative years on the north side. The clear implication is that I am doomed to be a Chicago Cubs fan. Perhaps only Boston Red Sox fans and the Dalai Lama can appreciate the internal state of resignation of interest in worldly success that this lifestyle induces. In 1989, I told all my friends in San Francisco that it would be *geologically* unsafe to violate the natural order of the universe by allowing the Cubs into any sort of playoff. Fortunately, the Cubs/Giants game produced only a terrifyingly destructive earthquake instead of, say, California sliding into a marine trench.

Nonetheless, following the Cubs predisposes me to think in terms of a universe in which puny humans struggle blindly against an implacable, pitiless, statistical fate. You may, in contrast, believe that individual heroics can sometimes save the day. Well, I'm the one writing the math book, I have an Ernie Banks autographed baseball, and I invite you at least to review some of my strange conclusions.

Baseball Fun

Baseball, probably because it is a slow-moving game, seems to inspire fans to study statistics. After all, fans have a lot of time on their hands even when the game is in progress. May as well study some numbers.

The problem with baseball statistics has to do with superstition. Very few people think that they need lucky charms to make the sun rise. It's more or less a certainty. But not a mile from my house is a Safeway that sells "Lucky Lottery" candles. The candle is in a big glass jar with numbers printed on the jar wrapper, which also lists detailed instructions for letting the candle help you pick numbers. From the appearance of the customers buying these things, the vendor is probably smart not to offer a warranty.

Who's hot, who's not: a coin-flipping experiment

These candles show two things: 1) People think a lot about random events, and 2) they will do all sorts of things to try to control random events.

The brilliant and innovative statistician Julian Simon once had an economics class pick stock portfolios from a set of "stocks" that he selected — he was trying to let the class see that short-term behavior of most stocks is fairly random. In this case, the stock prices were actually computer-generated random numbers. When he tried to stop the experiment after a few weeks, the students complained.

Several of the students were convinced that they had figured out a system to predict changes in the prices of the "stocks" and wanted to see how it panned out over a few weeks. Somewhere in the fundamental design of the human nervous system are both a belief in patterns and an absolute determination to find patterns whether they exist or not.

This brings me to the issue of slumps and streaks in baseball. I will now examine a month in the life of a .250 hitter, a fairly common species of baseball millionaire. Remember that .250 means that this player gets a hit one out of every four times at bat.

Take two coins and toss them. If they both come up heads, that's a hit. Any other combination will be "not a hit." Put your player somewhere in the starting lineup so that he gets somewhere between three and five at-bats per game. Play five games a week. That means

chances = 4 weeks × 5 games × 4 times per game = 80 chances

If you don't have the two coins handy, just look at Figure 18-1, where one round of the experiment has already been done. What you see is not someone getting a hit every fourth time at bat, as in Miss-Miss-Miss-Hit. You actually see something very much like a real one-month hitting record for a real player.

```
Baseball: Hits with a 0.250 average

           week 1     week 2     week 3     week 4

           Hit        no         no         no
           no         Hit        Hit        no
           no         no         Hit        Hit
           no         no         Hit        no
           no         no         no         no
           no         no         no         no
           no         no         no         no
           no         Hit        no         no
           Hit        Hit        no         no
           Hit        no         no         no
           no         Hit        no         no
           no         Hit        no         no
           Hit        no         no         no
           no         no         Hit        Hit
           no         no         no         no
           no         no         Hit        no
           Hit        no         Hit        Hit
           Hit        Hit        no         no
           Hit        no         no         no
           no         no         Hit        no

hits per week   7          6          7          3
```

Figure 18-1:
Two coins at the ballpark.

There are long dry spells in which your coins should be worrying about changing their stance. There are hot clusters during which the coins get a hit in every at-bat and tell sportswriters that they have finally found the secret to hitting pitches that are low and away. There are periods when the coins are having a routine day at bat (who knows, maybe the coins made a spectacular catch).

The point is that, in the analysis of hundreds of real and simulated batting records, *the average itself seems to be an index of talent, but the details of a season can't be distinguished from random events.*

Interpretive numerology

Starting with this one-month hitting record, consider the following bits of conversation:

> ✔ **Joe TwoCoin to Tim Slapshot of Channel 7 news:** "Well, Tim, you just have to give it 110 percent every day. I think this slump in the last part of the month was due to my changing shoes, but now that I'm back to the old shoes, I think things will turn around. . . ."

- ✔ **Six-Deer to Five-Rabbit, somewhere in Central Mexico, 1350 A.D.:** "You know, the last time we sacrificed 100 captives, we got *three days* of rain, and this time we get just one. What do you think . . . should we try 200 and see if the god Tlaloc is demanding a new rate of 100 captives per day?"

- ✔ **Bob Annuity, financial expert on *Money Talk AM:*** "We're all a little concerned about management practices at the Schmerdlap Fund. For five years there, they consistently outperformed the market indexes. But this is the third year in a row where they're slightly behind the averages. What gives? Are they losing their touch?"

Here's a principle that's perhaps less controversial than the one I would like to propose:

If you are following an activity that can be simulated by random events to the extent that you can't distinguish the simulated record from a real one, invest your time in something other than generating explanations. Maybe more captives will get you more rain, maybe not. Maybe you should have the captives work on the irrigation system instead.

"Percentage" Baseball

"But gee, Dr. Seiter, my distinguished guide to the world of numbers," you may ask, "if it's all so random, how come the good teams seem to win? In fact, what's random about the tendencies of your own beloved Chicago Cubs, who have managed to avoid the World Series in every year of your lifetime?"

Actually, the little experiment I just did is part of the answer to those questions. Two effects are at work:

- ✔ **Adding 'em up**

  A baseball team has nine players. On any given day, some of them are enduring a "slump," some of them are red hot, and some of them are batting their overall average for the day. A team whose batting average is .300 can be playing another team with a team batting average of .200 and losing. All it takes is the statistical coincidence of three or four of the players on the .200 team being "in the zone."

> **The odds**
>
> Of course, the .300 team is going to have more good days than the .200
> team. If you were willing to run two sets of nine experiments (you would
> want to generate random numbers on a computer rather than actually
> flipping coins and rolling dice), you would find, not all that remarkably,
> that the .300 team has about 50 percent more really good days than the
> .200 team. But the randomization effect means that you can go out to the
> ballpark and watch a team in the cellar beat the league leaders. That's why
> people keep showing up at Wrigley Field.

By the way, if you are one of those fans who follows baseball statistics closely,
you should try to find a copy of Earnshaw Cook's landmark study *Percentage
Baseball* (MIT Press, 1966). Better yet, send a copy to the manager of your
favorite team. Thirty years after Cook proved to a mathematical certainty that,
in practice, it's a waste of time to bunt to advance a runner on second, you still
see managers call for this play every few days or so.

The Truth about Wins and Losses

Here's a summary of the results of the baseball experiment:

Baseball teams win about half their games, plus or minus a small amount.

For sophisticated fans of European soccer, here's a restatement to cover your
own favorite sport:

Soccer teams win about half their games, plus or minus a small amount.

That's about it. In games that have low overall scores, it's harder for a given
team to dominate its opponents. When a typical score in a game is 1-0 or 2-1,
and each score is the result of several events with a big random component
added together, it strengthens the conclusion, "On any given day, a worse team
can beat a better one."

That's why lower-scoring games tend to be associated with fabulous salaries
and bidding wars for the best players. The contest can still be interesting even
if one team is wall-to-wall hotshots.

Higher-point games: Football

Because more points are scored in each game, a football team made up of the best individual players at each position would dominate its league more than a comparable baseball team. This conclusion is borne out by examining the records of the best and worst teams in all leagues in a given year. Look at Figure 18-2, a sort of cold-hearted analysis of the connection between typical scores in a game and the percentage spread in the sport between the best teams and the worst teams.

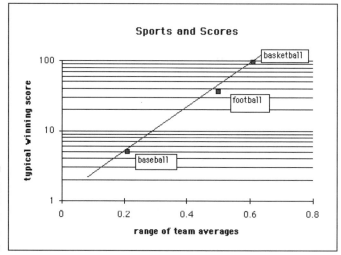

Figure 18-2: Points and spreads in football.

Higher-point games: Basketball

For basketball, the situation is even more extreme. A hand-picked team of the best basketball players could beat a team of hand-picked lesser players *every time*. Each game has lots of points, and each team has only five players, so it's easy to construct a totally dominant team even by picking players from the regular NBA rosters. Just think of an all-star team playing an "all-disappointment" team.

There's a lot of interest in preventing this situation. People have a tendency not to show up and watch games that one team has a 98 percent chance of winning. Thus, both basketball and football have elaborate rules for directing the best college players into the worst teams in any league. If there weren't special rules for redistributing talent in this way, especially in basketball, where a manager would only need to collect the 5 best players instead of the 9 or 11 best, you would see some teams finishing the season with 0.992 averages.

Betting on It

The essence of running a sports gambling operation is something like this: Formulate a bet that attracts equal amounts of money to either side, keep 5 – 20 percent of all the money bet, and pay off the rest to the winners.

As the distinguished Scottish analyst of sports betting, A. McAlpine, has pointed out, the situation reduces to a form of casino gambling in which you are offered a 50/50 bet at a fraction less than the expected even money odds. If your team beats the point spread on a $100 bet on the spread in a football game, you typically get back about $180. This is a better return than you get in state lotteries, but it's still a guaranteed way to lose money in the long run unless you personally have the resources to fix games.

 You will meet very few ordinary civilians who have made money betting on sports, which should tell you something. On the other hand, it's almost impossible to *lose* money offering illegal sports gambling services. That should tell you something, too. The side of the operation where you lend money to cover gambling debts at 50 percent interest per week is another intriguing aspect of a gambling service organization. That's about 17,000 percent interest over the course of a year, just as a mathematical curiosity.

 Before you embark on a career as a sports bettor, you may want to test your qualifications by proving that you can call the points on ten rolls of a pair of dice in a row.

Chapter 19

Statistics in the News

● ●

In This Chapter

▶ Distinguishing "good" medical studies from "bad" ones

▶ Understanding how scale relates to statistics

▶ Reading the news with a critical eye

▶ Understanding the facts about crime in the U.S.

● ●

*T*his chapter is about numbers that are reported to you every day on radio and television news and in newspapers and magazines. It is not, I regret to report, written from an entirely neutral point of view, at least as television news depicts neutrality.

My point of view, which is supported by the numbers in the stories themselves, is that very often there is much less than meets the eye. Because the news carriers operate 24 hours a day, 365 days a year, they have built-in mechanisms for generating news stories from data that often doesn't really merit reporting. Have you, for example, ever seen headlines such as these?

✔ **Three-Year Cholesterol Study Shows No Clear Results**

✔ **Most Crime Statistics Within Expected Bounds**

✔ **Police Work Safe Compared to Lumber Mill Jobs**

✔ **Stock Market Down for No Particular Reason**

✔ **Weather for Rest of Week Not Worth Reporting**

✔ **School System Doing About the Same**

✔ **Nearly All Gamblers Lose Money**

✔ **Most Small Commodities Investors Lose Their Shirts**

I don't think that you have. Nobody wants to tell you this stuff, and very few people want to hear it. For example, working the night shift in a convenience store is significantly more dangerous than doing nearly any form of police work short of SWAT team assault. That's simply a matter of reading the statistics. Nonetheless, nobody has managed to put together a highly rated television show about the dangers of convenience-store clerking (there's a great movie called *Clerks,* though), whereas the thrills of police work are dramatized several times a night.

Similarly, I can tell you from decades of personal experience at the frontiers of biotechnology that most research programs, even well-organized, well-funded programs inspired by brilliant ideas, don't often result in real medical breakthroughs. If you are diagnosed with lung cancer, for example, the recommended course of treatment won't be much different from the one used 20 years ago, and your life expectancy from time of diagnosis won't be vastly better. That's not much to show for the amount of research activity devoted to this topic. For that matter, we don't have anywhere nearly enough to show for decades of anti-smoking ads. There is, however, no shortage of stories about medical progress and fantastic developments that are just a few years away.

This chapter is not meant to discount in any way the efforts of millions of fine people trying to do their best every day. It's a complaint about the way numbers get distorted in news to make all sorts of events seem more significant than they really are. So this chapter explains the "resizing" of numbers in the news so that you can get a better understanding of the meaning behind those numbers. If you get interested in this sort of analysis, find a copy of the book *200% Of Nothing* by A. K. Dewdney (John Wiley & Sons, 1993), a detailed and hilarious study of statistics in the news. (Actually, anything by Dewdney is worth reading.)

Warning! Read the Label Carefully

I picked up the paper yesterday (*San Francisco Chronicle,* early February 1995) and read an alarming story about coffee. According to the story, reported on the AP newswire, a researcher in the Netherlands had discovered that swallowing small amounts of coffee grounds resulted in a significant increase in serum cholesterol. The testing program involved drinking lots of Turkish coffee, which is pretty much like regular old Italian-restaurant espresso only with a dab of mud in the bottom.

So does that mean that it's panic time? Well, on the face of it, one reason not to panic is that you probably don't get much coffee with grounds in your diet. Practically all the coffee in the U.S. gets dripped through a filter, so you may wonder why this exact bit of news, if that's what it is, deserved a page-one, inch-high headline. But I picked this study because it has all the numerical earmarks of studies that you should generally ignore. Here's the list.

A small sample

Reading the article, you find that the study involved about 24 university students who were divided into two groups of 12. Essentially, one group got the Turkish coffee, and the other didn't.

It's hard to be polite about this. You could hardly prove that dogs bark with a sample size this small. The statistical power of a test is related to its sample size. In correctly designed tests, bigger sample sizes allow researchers to make more definite conclusions. You can treat yourself to a great simplification in your medical-news intake with the following easy-to-remember rule:

Don't bother reading medical results obtained on fewer than one thousand test subjects.

If you start noticing sample size in articles, you'll be amazed at the number of dramatic results reported from studies by a single doctor with 17, 31, or 43 patients. You may miss a few worthwhile results by using the preceding rule, but you probably won't miss much, because studies tend to get repeated if they're paying off. In fact, if someone obtains a really promising result on a smaller sample, getting funding to repeat the study on a large scale is usually pretty easy. You can, in most cases, wait with impunity until a bigger study either confirms or contradicts the original study results.

No controls

A good study has very tight controls on conditions — that means, for example, that only one factor is different between the two groups being compared. As you read along in this study, it develops that the two groups were matched for starting cholesterol levels and that some attempt was made to control diet. But if one of the students went off on a coffee-inspired ice cream binge during the test, the results of the ice cream on a sample this small would be almost enough to account for the whole effect observed in the test.

Lots of studies, for example, are performed on people who are taking some form of medication as they go about their daily lives. Later, researchers count on these peoples' memories of what they ate, what they did, whether they took aspirin, or any number of other things. Sometimes these other factors matter, and sometimes they don't. But this lack of control over other possible influences is one of the reasons that you can also ignore most studies that begin, "A doctor in Philadelphia reports that ordinary baking soda can cure herpes," or the equivalent.

My point is this: *Always check to see whether the researcher has ruled out every other possible explanation of the results.*

With a little reading, you will soon find that most news stories actually report at least some of the conditions of a test, including the way the test was designed to control for various factors that might be important. If you think about it carefully, you can often figure out for yourself whether all the possible influences in the test were really controlled.

Consider the coffee "experiment." Statistically, if you take a small batch of students and test their cholesterol levels today, split them into two smaller batches, and test again two weeks later, the most likely result is that you will find some difference in the two batches. Does this mean that something special happened? Typically, it doesn't.

The Effects of Scale

You may come across a medical report that has a decent sample size and effective procedural controls but that makes no mention of the effect of dosage (amount of a drug, amount of a treatment, amount of peanut butter). The absence of this information is quite damaging to your ability to understand what's going on, because there are typically two kinds of scale effect. You need to know which kind of scale effect applies to the study at hand.

Straight goods

Figure 19-1 shows one type of scale effect. For twice as much exposure to some factor, you get twice as much effect.

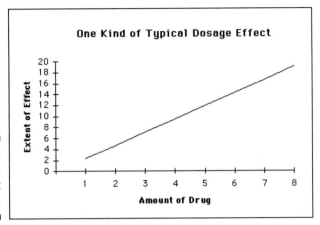

Figure 19-1:
A simple
dose-effect
relationship.

One example of this kind of relationship is ionizing radiation and DNA damage. Sometimes this is expressed as, "There is no safe amount of radiation." Every exposure to x-rays, therefore, produces some measurable amount of DNA damage. If you get ten times the exposure, you get ten times the damage.

When rescue workers went into Chernobyl to try to contain the radiation leakage, they weren't taking a risk per se. In the interests of trying to stop a plume that would take out a hundred thousand people, they were going to absorb an amount of radiation that was absolutely guaranteed to kill them. And they knew it. You can look up the dosage in millirems (a measure of radiation) in a table and read off the result: certain fatality. There was no chance that they could "get lucky" or that the radiation would miss or that somehow a stronger personal constitution could overcome the damage.

Some types of chemicals have this same relationship to liver damage. At any amount, you get damage, and at twice the amount, you get twice as much. Sniff enough petrochemical solvents, and you're gone.

Dosage curves

Most effects in human biochemistry aren't like this, however. For most substances, particularly the everyday substances most likely to be mentioned in news articles of the "Whole Wheat Bread Causes Insanity" type, the dosage curve looks like Figure 19-2.

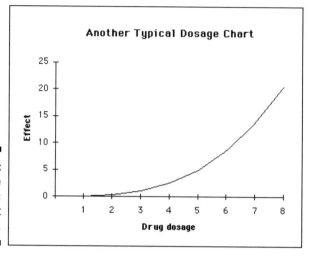

Figure 19-2:
A more
realistic
dose-effect
relationship.

What does this mean? Several things.

- If you drink a few glasses of wine per week, you'll be okay from a health perspective. If you drink a quart of Absolut Vodka every day for a few years, you're dead meat. It's not that drinking only 5 percent as much alcohol (as the quart-a-day schedule) has 5 percent of the effect — it doesn't have any *adverse* effect. The dosage effect is a curve, not a straight line.

- If you eat a half-pound of beef fat every day, you can acquire a whole variety of problems, one of which is rapidly inflating pants size (hey, that's a lot of extra calories). But eating small amounts of fat doesn't result in problems. Europeans visiting the United States are always startled to see what a collection of porkers we are, and grocery stores in Italy and France are not packed to the rafters with diet foods. You just eat much *less* of the same stuff, from potatoes to butter to beefsteak to ice cream, and it's all okay (as long as you get enough carbohydrates and fruits and vegetables).

- Remarkably, the health statistics for people who smoke only a few cigarettes (two to three) a day are barely distinguishable from the statistics for nonsmokers, especially in urban areas. In contrast, at three packs a day, you're up in the rapidly soaring danger zone of the curve, pretty certainly headed for a really ugly demise. The problem with this piece of information, and the main reason that medical authorities are reluctant to discuss it, is that experience shows that very, very few people hold on at the three-cigarettes-a-day level (and indeed, why bother?). But even cigarette damage obeys a curved dosage law rather than a straight line.

The conclusion from the two types of graphs is that you don't have the whole story until you know which type of graph is appropriate, and you don't know how to modify your behavior until you know the level at which some effect "kicks in." Badly calibrated results on different foods, for example, have made the U.S. a paradise for purveyors of goofy quack diets.

Crime

The good news is that you're not as likely to be shot by a criminal as you might think. The bad news is that virtually everything that comes out of your television is designed to keep you paralyzed with fear every day of your life. This section is intended to help you read crime statistics in newspapers (there isn't much in-depth statistical information on TV, I'm afraid) when you can find them. Of course, you are more likely to read accounts of individual lurid crimes than real statistics, but I will soldier on nonetheless.

To get the real facts about crime and statistics, go back to the bookstore where you bought this volume and ask them to order a copy of the college text *The Mythology Of Crime And Criminal Justice* by V. Kappeler, M. Blumberg, and G. Potter (Waveland Press [708-634-0081], 1993). It's not as exciting as a cop show on TV, but it's comforting bedtime reading.

Bang?

I used to live in Los Angeles, and I knew about five officers from the Los Angeles County Sheriff's Department personally. At a wedding reception one afternoon, I asked them how many shots they had fired on duty, in total (this was in the drug-crazed, crime-ridden late 1970s). The answer, to my surprise, was zero. As they explained, a long list of gun-handling rules have to be observed, and you can pretty much kiss your career good-bye if you violate them. The *typical* officer, not just in Anytown, Ohio, but in L.A., retires without anything resembling a gunfight on his or her record.

Contrast this, if you like, to the number of shots fired in a four-hour sampling in a single day on television. Day after day, year after year, the difference between real law enforcement on the streets and TV law enforcement gets wider (actually, the ancient *Dragnet* show was closer to reality), as pseudo-documentary TV shows scour the entire country trying to find instances of SWAT team live action to pep up the show.

Crimes of the century

One interesting and well-studied case of news crime statistics concerns serial murders. I'm just going to talk about serial murders here because there are several other statistical issues (incarceration rate, results of three-strikes-you're-out sentencing laws, or deterrent effects [or not] of the death penalty) that I wouldn't touch with an insulated fiberglass four-meter barge pole. After all, you should be able to read a math book without too many political hassles.

In the mid-1980s, many newspapers, repeating an original story that appeared in *The New York Times*, advanced the proposition that nearly 20 percent of all the murders committed in the U.S. were the work of serial killers. This number corresponded to about 4,000 murders, and the papers further suggested that this epidemic had exploded since the early 1970s. It was also suggested that half the victims were under 18 years in age and that the U.S. had about 40 such serial killers out wandering the U.S. at random, and pretty busily occupied at that. In the mid-1980s, the U.S. had about 5,500 unsolved murders each year, and a large fraction of these — nearly $2/3$ — were attributed to serial killers.

These figures bring up a number of questions. First, the annual murder rate overall has been fairly stable. It doesn't double from year to year. Second, the number of murderers identified as serial killers from 1950 to 1970 was exactly two, and one of them was the notorious Boston Strangler. Did something very strange happen in the year 1970 to produce a large crop of demented maniacs?

Nah. We've always had about the same number of demented maniacs as a proportion of the population. Actual Justice Department figures say three things: 1) There may be about 30 serial killers in the U.S. at large at any time, 2) they typically stick close to home, and 3) they kill four to six people per year. That means fewer than 200 serial killer victims per year, plenty bad enough, but not much like the figure 4,000. And it means that probably less than 1 percent of murders are committed by serial killers, not 20 percent.

So what was going on here? What was the basis of dozens of lurid TV mock-documentaries? The explanation has to do with the way crimes are reported in the Uniform Crime Reports (UCR). The early 1980s saw an upswing in the number of crimes first reported as both "motiveless" and "no suspect" and that were subsequently listed as unsolved. Looking at an assortment of cities and noticing which ones had a big upswing in "unsolved" murders (in a really typical murder, the spouse is still standing there in the kitchen holding the gun when the cops show up), there was in fact a powerful connection between unsolved murders and drug-trafficking activity. It happened that, in the mid-

1980s, most murders that involved finding the body of a known drug dealer in an alley, with neither drugs nor money nor witnesses about, was UCR-listed as motiveless and no-suspect. That means that a dead guy with half-a-dozen diamond rings and a matching number of small-caliber holes in his head, blown away inside a $200,000 sports car, was going on the books as "serial-killer victim," one more number to use to scare the daylights out of housewives in Joplin, Missouri.

The crude result of all this was that the original report of an upswing in UCR motiveless and no-suspect murders was directly translated into serial-killer murders in the widely publicized newspaper article, and that contents of that article were endlessly repeated. Any beat cop in Miami could have suggested that the dead bodies in the trunks of Ferraris were probably not the work of truly motiveless serial killers randomly attacking the public, but this perspective was never aired.

The math issue here is twofold. First, every time you read an article, you have to ask, "What is the basis of these numbers?" Second, it wouldn't hurt to ask whether the conclusions in the article violate common sense. This is pretty tough to do, since everyone's common sense is constantly being subverted by media distortions, but sometimes you can make a little investigation just on the numbers. In this case, the projected serial killers would have had to have been killing several people per week. Since there are no historical examples of real serial killers being able to maintain this rate (even Jack the Ripper averaged fewer than one per month or so), it is unlikely that some monster could be producing a few hundred dead bodies every few years in Cincinnati. Someone would notice.

Chapter 20
Puzzle Time!

- -

In This Chapter

▶ Decoding riddles in logic

▶ Substituting letters for numbers

▶ Solving those idiotic matchstick puzzles

▶ Finding the next number in a number series

▶ Solving algebra-type puzzles

- -

*F*or me, this chapter is a "problem chapter." It appears in this book as the result of a direct request from a colleague. He likes puzzles, and he consistently has a better idea of what should go in my books than I do. And somebody out there must love puzzles. Even on a modest grocery store magazine rack, you'll find volume after volume of logic and number puzzles by the publisher Penny Press (which appears to specialize in puzzles and crosswords). Evidence also shows that puzzles are good for you, especially if you are older — seniors who do puzzles regularly seem to avoid memory and other cognitive problems.

Personally, puzzles drive me nuts. Maybe it's because I am obliged to do plenty of computer mathematics in the course of a given week, but I never think of puzzles as having recreational purposes. Perhaps it's because of my horrible, no-fun perspective on things in general (when I see a puzzle, my first impulse is to write a short computer program to solve it), but the only game I can stand to play for any length of time is the Egyptian funerary game *Senet,* a sort of precursor to backgammon. Hey, it's got to be entertaining if your spirit is supposed to play it in a tomb for eternity.

So I'll try my best, but please bear with me. I'll analyze some typical kinds of recreational puzzles and share some strategies with you. If you're more interested in recreational mathematics than in the sort of math-first-aid basics that this book covers, go to a bookstore and find anything by Martin Gardner.

A Matter of Logic

This warm-up puzzle is like the type you may encounter on IQ tests and other standardized tests.

Read these two statements:

> No birds are insects.

and

> All eagles are birds.

Which of the following is true?

1) No insects are birds.

2) Some birds are not eagles.

3) No eagles are insects.

4) All birds are eagles.

There are several ways to look at this problem, from a full-tilt algebraic approach using the notation of formal logic to sheer guess-work.

Here's one of the simplest ways. If all eagles are birds, then birds make up a big grouping of flying things, and eagles are part of that grouping. If something is never true of a bird, it means it's also never true of an eagle. One of the two statements at the beginning is, "No birds are insects." So if a bird can't be an insect, it's also the case that no eagle can be an insect. So the answer is choice 3.

Because both beginning statements are true, it doesn't seem likely that together they could generate choice 4, which is not true. If you are ever taking a test with questions like this, and the test is being timed, one of the first things you can do is cross out choices that are obviously wrong. Doing so improves your chances by 33 percent!

Here's another example of a *test* puzzle:

Mr. Brown, Mr. Green, and Mr. White meet at their club. Mr. Brown remarks that between them they are wearing brown, green, and white outfits (I picture a contrast-stitched solid polyester leisure suit) but that no one is wearing his *name color.* The man wearing white agrees. Can you assign outfit colors based on this information?

The key to this problem is to figure out who is wearing white. You know that it's not Mr. White because of the *outfit color-exclusion rule.* You also know that it's not Mr. Brown because Mr. Brown made the first remark. That leaves Mr. Green, so you have Mr. Green in a white outfit. For the next step, you have Mr. Brown, who can't be wearing white and also can't be wearing brown. So Mr. Brown is wearing the green outfit — probably a neon avocado color with a monogrammed *B* on the pocket. The only case left is the silent Mr. White, fidgeting in embarrassment in his brown outfit. He is embarrassed, of course, because even though he exists only in a puzzle, he can see what color everyone is wearing, so the remarks strike him as terminally inane.

There are long puzzles with ten different people and conditions (Mr. Blue lives in a red house and hates dogs), but they all follow the same procedure — find the one case in the puzzle for which you can assign a definite characteristic and then hand out the leftover characteristics. Probably the most entertaining version of a logical puzzle ever developed is the board game *Clue* because it gives you an opportunity to brandish a tiny knife or candlestick while you rule out the logical possibilities.

Adding Letters?

Some of these puzzles are actually quite fiendish, but they can at least be solved by nothing more than trial and error.

Here's a puzzle recently used in a competition sponsored by *Games* magazine. This magazine, needless to say, is one of the best sources of difficult puzzles available.

```
    COCA
  + COLA
    SODA
```

In this kind of puzzle, you figure out which numbers to assign to letters to make the addition come out correctly. Where do you begin?

1. Well, you can look at the last bit of this puzzle and pretty much guess that $A = 0$; in most puzzles like this, finding zero is a good place to start. You have $A + A = A$, so A can't be anything else.

2. The next suspicious part of the pattern has to do with the letter O. You have already assigned the number zero, so that's not it. In the middle of this problem, you are probably carrying a one from the $C + L$ addition, and

you can figure that you are probably carrying another one from the $O + O$ part. That means O is greater than four, because otherwise there's nothing to carry.

3. Running through the choices for $O + O$ ($5 + 5, 6 + 6$, and so on), the only number that will work is 9. Adding $9 + 9 + 1$ gives you something that ends in 9, while adding $8 + 8 + 1$, for example, gives you something that ends in 7.

4. So now you have

```
   C9C0
+  C9L0
   S9D0
```

The choices for C are 1, 2, 3, or 4, because if C were any bigger, you would have something extra in the sum instead of plain old S. C can't be 4 because S would then be 9, and 9 is already taken. It can't be 1 because then $C + L$ wouldn't produce 1 to carry unless L were 9, and 9 is already assigned.

5. So the puzzle reduces to trying out 2 and 3 as values for C, and you find that 3 works. S becomes 7, D becomes 1, and L becomes 8.

This next one is an example of the division variation of this kind of puzzle.

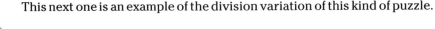

GGG/BGGFA = CF

Again, you are supposed to assign numbers to these letters. This particular example is one you should watch out for because it gives you some insight into the mind of the puzzle designer. Sometimes you see puzzles in which the letters spell out something cute, and sometimes you don't. As a first guess, if you don't see any letter farther along in the alphabet than J, see if the puzzle can be solved by either

$A = 0, B = 1, C = 2$ . . . and so on,

or

$A = 1, B = 2, C = 3$ with some other letter (usually X or Z) as zero.

If this doesn't work, your best bet in division problems is to see whether you can find 0 or 5 in one of the letters. Beyond that, you can usually look for the most likely candidate for 9.

Geometry Puzzles

Most of these puzzles aren't related to real geometry — they have to do with spatial visualization. Visualization in 3-D appears to be an ability that's not particularly easy to learn. Like musical ability or foreign language talent, there appears to be an actual correlation between various brain structures and visualization, so there's a lot of variation in geometry puzzle ability.

Look at the nine dots in Figure 20-1. The problem is to connect them with four lines without taking your pencil off the page.

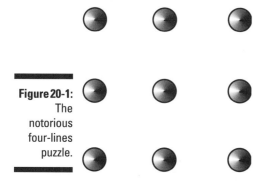

Figure 20-1:
The notorious four-lines puzzle.

The solution to this nine-dot puzzle is shown in Figure 20-3. In this, as in most other puzzles that take place in a plane or in space, the trick is to go *outside* the boundaries. When they first encounter this kind of puzzle, most people try to draw the lines *within* the set of dots.

Geometric imagination (this is a bit of a diversion, but please follow it anyway) has lots of applications. My father-in-law, Dennis Toth, was once working at a factory that had a huge spot-welding machine. The power electronics in this machine would break down about twice a year. When this happened, the standard procedure was to get two forklifts working together to back this huge block of equipment away from a wall so that the back access panel could be reached. The first time Mr. Toth saw this operation in dramatic industrial full swing, he opened a door next to the spot welder to see what was in the room behind the wall. The answer was *nothing.* In fact, the wall was just a drywall panel built to support a few electric plugs. The obvious maintenance step would be to cut a removable panel in the wall for access to the back of the machine. This hadn't occurred to anyone in the *12 years* since the welder had been put into place.

Suitably emboldened by that example of 3-D imagination, try this:

> I give you six matches. Your assignment is to find ways to make four triangles using all six matches at once.

Most of these match problems seem to have started out as bar tricks, usually with a bet placed on the outcome. This has the advantage that often the participants can't remember the answer from one weekend to the next! I'll give you a hint — this particular trick works better if the matches are arranged on a carpeted surface. The answer is shown at the end of this chapter.

Here's another silly match problem. Figure 20-2 shows a classical-style bank front made out of matches. The first problem is to move 2 matches in this pattern and make 11 squares. The second problem is to move 4 matches and get 15 squares.

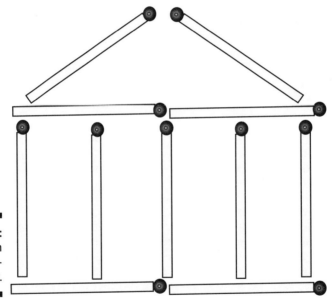

Figure 20-2:
More match rearrangement.

The answer, again, is shown at the end of this chapter. In this as in all other trick problems, you have to think a bit about what constitutes a square and how big the squares have to be. Just as in the connect-the-nine-dots puzzle, the solution is to go outside the pattern as presented; the trick here is to avoid letting the original layout dictate the scope of the arrangements you try.

Number Series

Lots of puzzles ask you to find the next number in a series. I'll just state four of these and then discuss some methods for solving them.

1) 2, 3, 4, 6, 8, 12, _?_

2) 46, 55, 64, 73, _?_

3) O, T, T, F, F, S, _?_

4) 1, 1, 2, 3, 5, 8, 13, _?_

The people who design number series (a string of numbers like this with commas should really be called a *sequence* in math, but in the puzzle world they always seem to be called *series*) for tests know in advance that you are going to be looking for something simple. By something simple, I mean series of the type

2, 4, 6, 8, 10, . . .

or

3, 6, 9, 12, . . .

in which you just add a term to the preceding number to get the next one, or

2, 4, 8, 16, . . .

1, 5, 25, 125, 625, . . .

in which you multiply a factor by the preceding number to get the next term.

Read on to see how a little psychology helps with these puzzles:

2, 3, 4, 6, 8, 12, _?_

Right away, you can determine that you don't add some fixed number to get the next term, and you are not multiplying some constant factor. That's the clue that you should look for a *double* series. Look at every other number in the list. The first set would be 2, 4, 8, and the second would be 3, 6, 12. That means that there is an original *pair* of numbers, 2 and 3, at the start of the series. Multiply these by 2, and you get the next two numbers, 4 and 6. Multiply 4 and 6 by 2, and you get the next pair. Can you take it from here?

46, 55, 64, 73, _?_

What's going on here? It's a case in which there are actually two ways to see the pattern. As a straight number-addition series, each one in this list is just nine more than the number in front of it.

You can also see that the first number in each pair of digits is going up by one, and the second digit is going down by one. Of course, that's just what happens when you add nine. But often, series of three-digit numbers are constructed in which each digit of the three-digit number follows its own little rule.

O, T, T, F, F, S, _?_

I know, I know, this is really stupid. But this trick turns up a lot on popular quizzes, IQ tests, and everywhere else. After reading this chapter, you should be able to knock it off at a glance. Write out the words one, two, three, four, and so on, and look at the first letters of the words. That's the story.

One odd variation of this kind of series that I saw on a Mensa test was

T, O, F, O, F, N, T, S, F . . .

The next letter is T. This series represents the word/number conversion of the numbers in the representation of π. (Read out "three one four one five nine . . . ") Sometimes people are just too cute for their own good.

1, 1, 2, 3, 5, 8, 13, _?_

This series isn't just a goof like the rest of them. This is real mathematics. To find the next number, you are supposed to notice that each number is the sum of the two numbers before it. This is called the *Fibonacci series* (Leonardo Fibonacci wrote about it in 1228), and the numbers can be shown to have all sorts of interesting properties. They turn up in problems in physics, computer science, and the mathematics of compound interest.

Algebra Puzzles

Yet another kind of puzzle is really just an algebra problem. Look at this one, very slightly paraphrased from a real algebra textbook.

Binky the Cat is three times as old as Stinky the Cat. In five years, Stinky will be twice as old as Binky is now. How old are these critters?

You solve this problem by assigning quantities and doing the algebra. If Binky's age is *B* and Stinky's is *S,* you can translate the problem to

$B = 3S$

$S + 5 = 2B$

Plugging the first expression into the second, you get

$S + 5 = 2(3S)$

$S + 5 = 6S$

$5 = 5S$, so $S = 1$.

Thus Stinky is an adorable one-year-old kitten, while Binky is a mature cat of three.

Now the machinery of algebra as taught in high school is perfectly adequate for doing all these dippy "Gladys is ten years older than Martha, whose age is the square root of 81" problems.

My complaint is that most people are badly equipped to solve this puzzle:

Fred finances a $20,000 Camaro by refinancing his house at 9.25 percent. Ted continues to drive his paid-off car and puts the few hundred dollars once used for the car payments into paying principal on his home loan each month. What are their relative financial positions after five years?

I am distressed to report that many of the same people who can do match tricks and the O, T, T, F series with the greatest of ease don't have a clue about where to begin on this last puzzle. It is solved in Chapter 11 of this book, along with a bunch of comparable conundrums.

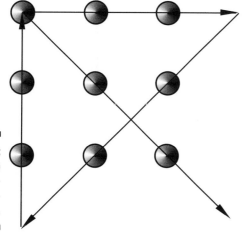

Figure 20-3:
Solution to connect-the-nine-dots puzzle.

Figure 20-4:
Solution to
the 3-D
match
puzzle.

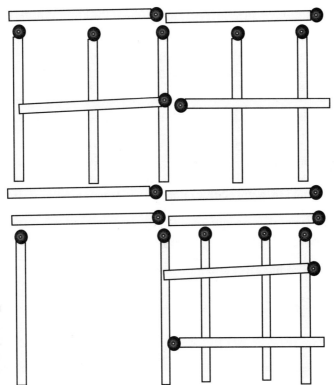

Figure 20-5:
The bank
match
puzzle
solutions.

Part V
The Part of Tens

"I SUPPOSE THIS ALL HAS SOMETHING TO DO WITH THE NEW MATH."

In this part...

This part gives you lots o' lists: ten stupid math tricks, ten calculator tips, and more.

Chapter 21

Ten Fast, Useful Math Tricks

● ●

In This Chapter

▶ Store math: Rounding and grouping prices

▶ Doubling and interest rates

▶ Checking your answers with the nines method

▶ Squaring numbers

▶ Multiplying not-square numbers

▶ Dealing with high finance on a calculator

▶ Converting and other metric matters

▶ Checking by round-off

● ●

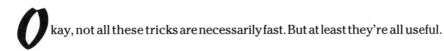

*O*kay, not all these tricks are necessarily fast. But at least they're all useful.

This chapter, in my opinion, distinguishes *Everyday Math For Dummies* from most other works on math. Needless to say, when I was first contacted about writing this book, I went out and bought everything even remotely comparable on the market. These books contain, among other things, tips on rapidly dividing numbers by 35 in your head, all faithfully copied by one author after the other from Gerard Kelly's *Short-Cut Math* (Dover, 1969), which at $2.95 is still probably the book to get if you want to pretend that calculators don't exist.

But some of these calculating "tips" in rapid-math books really provoke the remark, "Huh?" The tip for rapidly multiplying a number by 12, listed in one such book, is as follows:

1. **Multiply the number by ten.**

2. **Double the number.**

3. **Then add these two intermediate results together.**

The author doesn't seem to have noticed that *this is just the definition of multiplying by 12!* It's not a tip, it's not a trick — it's exactly what they told you to do in school. What I'm going to try to do in this chapter is at least to tell you something that you may not already know.

Store Math, Part One: Rounding

The people who run stores firmly believe that you perceive $17.99 to be significantly less than $18 — careful studies show that you tend to round the number *down* instead of up.

Of course, stores also use an elaborately calibrated class-system of price tag identification, ranging from fluorescent orange tags that say $7.00 to discreet little brass numbers that say 800 (no dollar sign). Sometime, take a field trip to some downtown store windows in a big city and watch how the stores signal their position in the retailing food chain with price tag conventions (besides just prices!).

Anyway, through the vast middle reaches of pricing, all this stuff about prices ending in nines reigns supreme. So here's a method for adding prices that also works with other numbers that can be rounded up or down easily.

Suppose the problem is to add these numbers:

```
   13.99
    7.98
   21.99
 + 4.97
```

The last thing you want to do is spend five minutes carrying ones and so forth. So rewrite this in two columns:

```
   14.00      1
    8.00      2
   22.00      1
 + 5.00       3
```

What's going on here? To get the values in the second column, you just take the tiny difference between the starting number and a nice, even number with 00 at the end. You can think of the second column as a sort of correction factor.

Here's how you process the answer. First, add the 8 and 22 to get 30. Add 5 to the 30 to get 35. Then add 14 to the 35 and you have 49. So the total would be $49, except for the column of correction factors in pennies. 1 + 2 + 1 + 3 = 7, so the total of the original numbers is

$$\$49.00 - \$0.07 = \$48.93$$

Officially, you have done more mental operations by breaking up the problem this way, but they were all easier than the original addition would have been.

Store Math, Part Two: Grouping

In the preceding example, I slipped a little mini-trick past you in the addition. That is, I added 22 and 8 first instead of just doing things the textbook way. You will find very few addition problems in which judicious pairing of numbers for adding doesn't simplify things marvelously. That's especially true if the numbers have been conveniently rounded off first. Here's a nice, long addition problem. As one way to prove that this method works in general, I had a computer select all these numbers at random:

17
151
23
24
205
39
75
66

First, look over the column and find all the pairs that make a sum ending in zero. If you add up all these pairs, you reduce the problem to

17	(17 + 23)	40
151		
23	(151 + 39)	190
24		
205	(205 + 75)	280
39		
75	(66 + 24)	90
66		

This third column doesn't have obvious groupings at this step, but you take (90 + 190) first to get 280 and then take (40 + 280) to get 320. *These* two values add up to the nice, round number 280 + 320 = 600.

The other advantage of doing this type of problem in pairing steps, even if you don't have absolutely perfectly convenient numbers, is that the answer is easier to check. You never add more than two numbers at a time, which puts less strain on your abilities, and you can check these two-at-a-time additions at a glance.

Playing Doubles: Interest Rates and Doubling Time

This section gives you just a little table relating interest rates to doubling time. The reason you may need to know this is that someday you may want to retire, which is going to take a considerable amount of money.

You'll need lots of money because of your fabulous extended life expectancy. When retirement age was first set at 65 in the Prussian State Civil Service in the late 1800s, the thrifty Prussians expected retirees to die within a few years of retirement, and the mortality tables confirmed this expectation. It was a simple business — you work for 45 years, save a little each year, and then spend it in two years of retirement and die before you have to see World War I.

You, on the other hand, are planning to work from age 21 or so until about age 65 and then live for about 20 more years. How much do you figure that you should save to work for 40 years and then retire for 20? Uh-oh. The Prussian arithmetic isn't going to work this time.

One simple way to calculate what your savings will be worth at some point in the future is to consider doubling times. Table 21-1 is a condensed financial table showing how long it takes for your money to double at different annual interest rates:

Table 21-1	Doubling Time versus Interest Rate
Annual Interest Rate	*Years to Double*
5%	14.2 years
7%	10.2 years
9%	8 years
11%	6.6 years

The problem, of course, is finding a reasonably safe investment that has a real return (adjusted for inflation) that's high enough — 9 percent or higher would be nice. That's beyond the first problem, which is collecting some money to invest in the first place.

Doubling for Multiplication

As long as the subject of doubling has come up, I may as well tell you about a doubling trick for multiplying numbers. Sometimes this method is called *Russian Peasant Multiplication,* but, in fact, it was apparently known 4,000 years ago.

The procedure is based on the observation that any number can be made up of powers of the number two — that is, numbers made of multiplying two by itself. Look at this list of powers of two:

- ✓ 1
- ✓ 2
- ✓ 4
- ✓ 8
- ✓ 16
- ✓ 32
- ✓ 64
- ✓ 128

Any number less than 128 can be made up as a sum of those numbers from 1 to 64. Here are some examples:

$$43 = 32 + 8 + 2 + 1$$

$$55 = 32 + 16 + 4 + 2 + 1$$

$$17 = 16 + 1$$

So if you want to multiply 195 by 17, you can multiply it by 16 and then by 1 and then add the two results together. The only reason this rigmarole is worth any attention is that you can multiply a number by 16 or 32 or 64 and so forth just by doubling that number a certain number of times. The scheme for multiplying 195 by 17 goes like this:

$$1 \rightarrow 195$$
$$2 \rightarrow 390 \text{ (two times 195)}$$
$$4 \rightarrow 780 \text{ (two times 390)}$$
$$8 \rightarrow 1{,}560 \text{ (two times 780)}$$
$$16 \rightarrow 3{,}120 \text{ (two times 1,560)}$$
$$1 \rightarrow 195$$

and then adding

$$\begin{array}{r} 3{,}120 \\ + 195 \\ \hline 3{,}315 \end{array}$$

Again, the advantage of this method is that it's easy to check the steps. I also bring up this method here because it is essentially the way computers multiply numbers — the computer converts all the numbers into sums of powers of two and converts them back into familiar decimal numbers only when *you* have to look at them.

Using Nines to Check Your Answer

An old method for checking correctness of an answer to a multiplication or division problem is still worth knowing, in case you have to work without a calculator. The method is based on so-called "9s remainders," an ancient trick for checking calculations. To get a 9s remainder, you add the digits in a number and keep adding until you have a single digit. And if the digit is nine, the 9s remainder is said to be zero.

I'll just do some examples rather than give you a rule:

✔ **756:**

Add the digits $7 + 5 + 6 = 18$.

Add these digits $1 + 8 = 9$.

9s remainder = 0.

✔ **631:**

Add the digits 6 + 3 + 1 = 10.

Add the digits 1 + 0 =1.

9s remainder = 1.

✔ **175:**

Add the digits 1 + 7 + 5 = 13.

Add the digits 1 + 3 = 4.

9s remainder = 4.

To check a multiplication, you find the 9s remainder for the numbers you are multiplying, and the product of the two numbers should have the same 9s remainder as these two remainders multiplied by each other. This isn't as bad as it sounds. Look at this case:

$$\begin{array}{r} 1438 \\ \times\ \ 615 \\ \hline 7{,}190 \end{array}$$

$$\begin{array}{r} 1{,}438 \\ \times\ 8{,}628 \\ \hline 884{,}370 \end{array}$$

Now, the 9s remainder of 1,438 is 1 + 4 + 3 + 8 = 16, and 1 + 6 = 7.

The 9s remainder of 615 is 6 + 1 + 5 = 12, and 1 + 2 = 3.

Multiply these to get $3 \times 7 = 21$; the check for this answer is 2 + 1 = 3.

The rule says that you should then get 3 as the 9s remainder for the product 884,370:

8 + 8 + 4 + 3 + 7 + 0 = 30, 3 + 0 = 3

This seems like a bit of work, but it's a better alternative than just stepping through the multiplication again, where you are likely to make the same mistake you made the first time.

Squaring Numbers

Numbers ending in zero are easy to square. You just square the first digit and tack on a pair of zeroes. That is

$$30 \times 30 = 900$$

$$40 \times 40 = 1{,}600$$

$$70 \times 70 = 4{,}900$$

and so forth. Knowing a few squares can't hurt, especially since you get them for free when you memorize your times tables. The information comes in handy when you're deciding how many overpriced hand-made tiles you need for your kitchen floor, for example.

What happens if you have a number that's almost, but not quite, easy to square? Suppose you have a number like 41 instead. For 41, think to yourself 1,600 (the square of 40), add 80 (2×40) to this number to get 1,680, and then add 1 to get 1,681.

You're just following this rule from algebra:

$$(D + d)^2 = D^2 + 2Dd + d^2$$

In the case of 41, you're saying that 40 is the big D and 1 is the little d, and then you're plugging them into this simple expression. It looks like this:

$$(40 + 1)^2 = 40^2 + 2 \times 40 \times 1 + 1^2$$

$$(41)^2 = 1{,}600 + 80 + 1$$

Multiplying Almost-Squares

A variation of the method I just showed you for multiplying numbers uses a very similar expression. To multiply 39 by 41, first think to yourself 1,600 and then subtract 1 to get 1,599.

Wow! That was simple! What's going on? Well, the expression from algebra is

$$(D + d)(D - d) = D^2 - d^2$$

Once again, $D = 40$ and $d = 1$. Can you see how to get a figure for 38×42 by using the same formula? If you came up with 1,596, good for you!

High Finance

Let me be blunt. Of all things you have ever spent $15 dollars on (and one of them could have been a small microwaved pizza in O'Hare Airport), one of them should have been a financial calculator. Because they often have names like MBA-100 and so forth, there's a misconception that you should be a high-powered financial type to own one. That's not true. If you ever had a credit card, you should own one.

Now, after you get your financial calculator, you will probably lose the little manual, so I'm going to review what the main keys mean. Photocopy this page about three times and file it in obscure desk drawers so that you'll always have a copy.

The main keys are the following:

✔ N

✔ PV

✔ FV

✔ %i

✔ PMT

N means *number of periods.* In a five-year auto loan, *N* is 60 because interest is applied monthly and $5 \times 12 = 60$. For a 30-year mortgage, the number is 360. Often people forget and just enter the number of years, which usually is incorrect.

PV stands for *present value,* but what it means for you is the amount you are borrowing. If you have an $11,000 car loan, *PV* is 11,000. If you have a $135,000 mortgage, *PV* is 135,000. Because most of the time you are figuring payments rather than calculating how much money you need to deposit to have some future value, this is the most appropriate working definition of *PV.*

FV is *future value.* You can enter a future value, an interest rate, and a number of periods and find out how much money you would need to deposit to end up with that *FV* after all the periods of interest. But you usually have the calculator figure out *FV* rather than entering the values yourself.

%i is *interest,* but it's the interest *per period.* That means that for most kinds of payments, you have to divide by 12 to get the number to use. An additional problem is that some calculators expect you to enter the number **8** if the interest rate is 8 percent, and others expect you to enter **0.08**.

Here's how to check. Enter **15** for *%i,* **1** for *N,* and **100** for *PV.* Have the calculator compute *FV* (there's a key labeled CPT or Compute or something similar). If you get the number 115, then you can just enter the interest as the percent number — if the problem says 6%, you enter the number **6**. If you get 1,600 for *FV,* then you should have entered **0.15** — the calculator wants you to enter **0.06** here if the problem says 6%.

And in problems where the interest is computed monthly (360 = *N* for the 30-year mortgage), you divide the interest rate (the 7% to 9% *annual* rate) by 12 before you assign it to *%i.*

PMT is *payment.* After you enter the other stuff, usually *%i, N,* and *PV,* you have the calculator compute the payment. If the payment looks about right, you entered the right values for *%i* and *N.* When you forget to divide by 12 on a monthly interest problem, the answers are *way* off, not just a little.

Metric Fun

Living as we do on this giant continent with miles of salt water separating us from everyone else, we get to use our own amusing medieval units for things. From time to time, however, you are obliged to convert things to units used by the rest of the planet. All you have to do to perform the conversion correctly is get the factor pointed the right way.

I'm going to try a little memory device, first for kilometers and then for kilograms. Just stare at this box for a few seconds:

MILES	kilometers

The point that I'm trying to make is that miles are bigger than kilometers. When you do a conversion, you end up with more kilometers than miles. The factor involved is 1.6. Now I'll explain what this means:

1) You have 138 miles. What's that in kilometers? You have to get more kilometers than miles, so you multiply by the factor

$$138 \text{ miles} \times 1.6 = 220.8 \text{ kilometers}$$

2) The sign outside Paris says 115 km (that's kilometers). How far away are you in *real* distance, Tex? The number for miles has to come out smaller, so you *divide* by 1.6.

> 115 kilometers ÷ 1.6 = 71.9 miles

With your foot flat on the accelerator of a Citroën-Maserati SM, you'll be there in a half hour!

Now just stare at this second box for a bit:

KILOGRAMS	pounds

Kilograms are bigger than pounds. When you do a conversion, you end up with more pounds than kilograms. The factor involved is 2.2. Now try some more conversions:

1) You weigh 165 pounds. What's that in kilograms? You have to have more pounds than kilograms, so you divide by the factor

> 165 pounds ÷ 2.2 = 75 kilograms

2) Your grandmother's German cookbook calls for 1.5 kilograms of bread chunks to make goose stuffing (remember to keep the schmaltz). What does this mean in an American grocery store? The number for pounds has to be bigger, so you *multiply* by 2.2.

> 1.5 kilograms × 2.2 = 3.3 pounds

Figure slightly less than three and a half pounds (recipes are never exact, certainly not for stuffing). At any rate, you'll get the direction of the conversion right if you can remember the size of the type in those gray boxes.

Rounding Numbers

Even if you use a calculator, you can mis-enter numbers. Probably the best way to check any arithmetic problem is to use rounded off numbers to make sure that you have the answer approximately right.

Take the example from the 9s remainder case earlier in this chapter. The 9s remainder is a good way to check against odd mistakes, such as remembering 9 × 9 as 83. But it leaves open the possibility of other errors that accidentally give you the same 9s remainder.

In the round-off check version of this problem:

$$\begin{array}{r} 1{,}438 \\ \times\ \ \ 615 \\ \hline 884{,}370 \end{array}$$

I would take the first number as 1,500 and the second number as 600. That gives me four zeroes to account for in the end, and the multiplication of 15×6.

15×6 is just 90, and when I tack on the four zeroes I have as my rounded-off answer:

90 0000 = 900,000

Actually, this is within 2 percent of the correct answer. This problem is more complicated, in fact, than most. I would bother to keep two digits in one of the numbers in the multiplication only if the first digit is one. On a problem like the following:

$$\begin{array}{r} 4{,}458 \\ \times 3{,}129 \end{array}$$

I would just take the 4 and the 3 and try

4,458 approximates to 4,000

3,129 approximates to 3,000

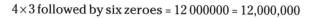

4×3 followed by six zeroes = 12 000000 = 12,000,000

You can go through life and be pretty satisfied with your calculations if you always have the right number of decimal places.

Chapter 22
Ten Fast, Silly Number Tricks

. . . .

In This Chapter

▶ The age trick
▶ The birthday trick
▶ A kid trick
▶ Grain and the pharaohs
▶ Almost a puzzle
▶ Fiscal humor
▶ A dicey business
▶ Mental miracles
▶ The Afghani birthday problem
▶ More fabulous powers

. . . .

*I*t's highly appropriate that there's a chapter like this in this book, even though as a professional grump I of course disapprove of all frivolity. The fact is, all these tens lists in the . . . *For Dummies* books started as a sort of homage to the lists on David Letterman's show. And now, right here in *Everyday Math For Dummies,* you have a Stupid Math Tricks section. What's next?

The Age Trick

This goofy party trick assumes that your friends can at least add and subtract a bit. Ask someone to multiply his or her age by 10. Then have that person subtract 9 times any number from 1 through 9, just to make it challenging. Ask for the number.

You add the last digit of the number to the first two to get the age.

Let's step through this: Suppose Steve's age is 25. He multiplies it by 10 and gets 250. Suppose he subtracts 45 (5 × 9) from this to get 205.

You hear the number 205 and partition it into 20 and 5 and add the two numbers together, giving you 25. You got it!

Suppose he had picked some other multiple of 9. Just to be complicated, suppose he picked 81 (9 × 9). That gives him

$$250 - 81 = 169$$

as the key number. You separate it into 16 and 9, add them together, and announce the result: 25!

What's going on here? Actually, the trick is very similar to the method of "casting out nines" to check on any addition or subtraction. Note that the trick works with *two-digit* ages — it fails for six-year-olds and for people over 100. On the other hand, people under ten and over 100 are very rarely reluctant to state their ages. I once went to a party where I was persistently reminded by a young lady that she had just turned four a few days earlier, and I was accosted several times by an elderly gentleman who reminded me, shaking his head as if in disbelief, that he was more than a century old. "More than a century . . . can you imagine that, Sonny?" The guy is now 105!

The Birthday Problem

The birthday problem is a standard in probability theory, but it has wide application. The question is how many people do you have to have in a room before it becomes likely that two people have the same birthday? Now, please note that this is very different from trying to calculate the odds on someone showing up with a particular given birthday, say April 26. Here's how you work this problem:

1. You have one person in the room. A second person walks in. The second person has a 364/365 chance (0.997 — that's a 99.7 percent chance) of *not* having the same birthday.

2. Now a third person walks in. This person has to miss *two* days of the year to not have the same birthday as either of the others. The odds on that are 363 in 365 (0.995).

3. Now a fourth person walks in. This person's birthday has to miss three other days of the year. Those odds are 362 in 365 (0.992). That is, the person has 362 chances out of the 365 days to *miss* the other birthdays.

4. The chances of everyone *not* having the same birthday can just be calculated as

Probability that all birthdays miss each other =

$(364/365) \times (363/365) \times (362/365) \times (361/365) \times (360/365) \times (359/365) \times \ldots$

$= (0.997) \times (0.995) \times (0.992) \times \ldots$

$= (0.989) \times (0.986) \times (0.984) \times \ldots$

All these numbers are pretty close to 1, but you can convince yourself that if you multiply enough numbers like 0.96 together, you rapidly get a pretty small number. Get out your calculator and multiply 0.96 by itself ten times!

5. The big product of all these 0.99-type numbers falls to less than 0.5 when you multiply 26 of them together. To be on the safe side (building a few more numbers into the factor), you can predict that two people will have the same birthday when there are 30 or so of them in a room.

People find this result surprising, but then coincidences catch people off guard. You find yourself seated on an airplane next to someone who was in your American History class in tenth grade, and both of you are knocked out by the coincidence. But the chances of something like this never happening to you are actually fairly small. As in the birthday problem, you are saying that *some* coincidence is likely, but you don't know which coincidence it will be (you don't know what the magic birthday will be).

Pennies from Heaven

That last section was actually an outburst of real math, albeit applied to a nonlethal problem.

Ask someone this question (preferably someone under the age of nine):

Why are 1995 pennies, in good condition, worth almost $20?

The answer is that 1,995 pennies, all piled up, are only 5 cents short of being 2,000 pennies. This is almost the numerical equivalent of a bad pun.

The Latest from 1600 B.C.

If each of seven houses has seven cats, and the seven cats are each served by seven mice, and each of the seven mice is hoarding seven ears of grain, and each grain has seven kernels, how many kernels of grain do you have?

$$7 \times 7 \times 7 \times 7 \times 7 \times 7 \times 7 = 823{,}543$$

I just threw this in here because of its outstanding formal resemblance to the nursery rhyme: "As I was going to St. Ives, I met a man with seven wives...." It comes from the Rhind Papyrus, an Egyptian mathematics text for training scribes.

Switching Digits

Figure out which two digits need to be switched to make this multiplication correct:

```
        324
    ×    68
      2,992
    × 2,744
     25,432
```

You can write this down on a scrap of paper and leave it around an office for someone else to find. It has been used to drive people crazy for several decades. Actually, problems of this type are fairly easy to construct if you want to make a career out of being obnoxious.

Money, Money

Offer to pay a hundred dollars to anyone who can give you five dollars in ten coins. The coins have to be half-dollars, quarters, and dimes, and you have to use at least one of each. Amazingly, people will spend several minutes counting change and making little calculations on this, and some never give up.

If the person could give you ten half-dollars, that would be $5. The requirement that you use other coins makes it impossible.

Dice Trick

This is essentially a math trick because it's based on addition and subtraction. You hand someone three dice and tell him or her to stack them up. There are now five *hidden* faces, meaning the parts of the dice that can't be seen. You ask your friend to check the total of the dots on the hidden faces while your back is turned.

When you turn around, you simply notice the number of dots on the face on the top of the stack. After much grimacing and mock concentration, you mentally subtract this number from 21 and announce the result.

The story is that the number of dots on the opposite faces of each dice add up to seven. For three dice, the total number of dots on the opposite faces is 21. In this case, the top face is showing, so you subtract the number of dots on the top face from 21.

Speed Arithmetic

Pick two single-digit numbers. For this example, use 3 and 7. Then write them like so, and make a series of numbers underneath them by adding them in pairs until you have ten numbers. It looks like this:

- 3
- 7
- 10
- 17
- 27
- 44
- 71
- 115
- 186
- 301

Because you have read *Everyday Math For Dummies,* you can now do fabulous calculations in your head. You claim that you can add these numbers in your head at a glance. The trick here is to multiply the fourth number from the bottom by 11. It's 71, and 71 times 11 is 781 ($71 \times 11 = 71 \times 10 + 71 = 781$).

You may have noticed that except for a few tricks in the previous section, there isn't much mental arithmetic in this book. Looking over the earlier literature on mental arithmetic from 1900 to the present and evaluating it, I concluded that a trick for rapidly multiplying large numbers by 17 (of course there's a special trick for it) is not especially helpful in daily life. If I could have you remember just a few things a year from now, I would rather that it be how to make compound interest work for you rather than against you. Even if you never learn tricks for doing long division in your head, you should have a bad gut feeling about financing ski vacations on 22 percent credit cards.

Afghani Birthdays

This is a variation on the preceding birthday problem. You are at a party in a large city, and some of the people at the party are from Pakistan, Afghanistan, Iran, and other Islamic countries.

You propose that you cannot only *see* that some of the people have the same birthday, but you propose that you *know* the birthday. It's January 1! Wow, all these people are Capricorns!

What happens is that in many societies, people just don't record actual birthdays, especially in small rural villages. When people from those societies get passports to come to the United States, the birthday is usually listed as January 1. In the hardball version of this trick, you size up the party and tell some unassuming American friend that it's a mathematical certainty that at least two people at the party have a birthday on January 1. The people with the arbitrarily dated passports are, of course, automatically in on the joke because they know that the date was assigned in an office.

Our focus on birthdays in the West probably comes from the association of the calendar with saints' days and the practice through most of European history of naming Christian babies after saints. Actually, from a practical point of view, knowing your own age to the exact day becomes pretty irrelevant past the age of seven or so, unless you live in a society like ours that sets great store by exact dates.

The All-Seeing Mind

This is a calculator trick. Have someone perform these steps on a calculator, and claim to be following mentally:

1. **Enter any three-digit number (the fabulous secret number).**

2. **Multiply the number by 10.**

3. **Subtract the secret number.**

4. **Divide this result by the secret number.**

5. **Square this number.**

6. **Add 19.**

After appropriate theatrics, you grandly announce the result: 100. What's really going on here is that the first four steps simply turn any number into 9 (effectively, you are dividing the number into 9 times itself). The square of 9 is 81, plus 19 is 100. But if you have a *victim* for this trick who's truly hopeless at arithmetic, he or she won't see the gimmick.

Chapter 23
Ten (or So) Numbers to Remember

O kay, you don't have to remember all of them. Some of these numbers are useful, and some are just for party chit-chat or to prove that you have an education. What is remarkable is the way individual numbers can have long stories, historical associations, and odd properties.

Pi (π)

Recall my discussion of geometry, if you will. Pi, or π, is the ratio of the distance around the edge of the circle (the *circumference*) to the distance across the middle (the *diameter*). In other words,

π = circumference/diameter

Therefore

circumference = π × diameter

Although you are taught in school that $^{22}/_7$ (or $3 + ^1/_7$) is a "good enough" approximation to use, just the number 3 isn't so bad if you have trouble remembering fractions (3 appears to be the value recommended in the Old Testament). If you have a circle that's 1 foot across, using 3 for π gives you

circumference $= 3 \times 12$ inches (1 foot) $= 36$ inches (1 yard)

Using the eighth-grade school value of $^{22}/_7$, you get

circumference $= ^{22}/_7 \times 12$ inches $= 37.7143$ inches

The exact circumference (using a better approximation for π) would be 37.6992 inches (to just four decimal places). For a geometry adventure of this size, the difference between $^{22}/_7$ and real π is 0.015 inches — less than a sixtieth of an inch. That's actually less than the width of the ink on the lines of a yardstick.

One common value of π in Egyptian practice was 3.16, which, since it's correct to 1 percent, probably didn't produce any errors worth arguing over on the banks of the Nile. The first really serious calculation of π was probably due to Archimedes, who used many-sided regular polygons to make an estimate of approximately 3.1418. By later times in the Greek world, the number 3.1416 was in common use.

In China, the number 3.162..., or the square root of 10, was a popular choice for π. It was recognized that the real value was closer to 3.14, but because 10 is a "perfect" number with lots of symbolism, the Chinese had a romantic attachment to its square root. By 400 A.D. or so, Chinese astronomers had worked out $^{355}/_{113}$ as an approximation. This fraction is 3.1415929204; when you compare it to the real value of 3.1415926535..., you may realize that it would be very difficult to determine this number by measurement since it's correct to a few parts per million and no one in fourth-century China had instruments of that accuracy. How was this done? Beats me.

Avogadro's Number

This number connects the world of individual atoms with the world of amounts of materials you can actually see. For short, the value is just 6.02×10^{23}.

Here's a rough explanation. Hydrogen is the lightest element, with an atomic weight of 1. Oxygen has an atomic weight of 16. A water molecule (H_2O) has 2 hydrogens and 1 oxygen, for a *molecular* weight of 18 $(16 + 1 + 1 = 18)$. A molecular weight of 18, in turn, means that 18 grams of water is 6.02×10^{23}, Avogadro's number, of water molecules. This quantity is called a *mole*.

The main point here is that, because atoms are very little, this number is very big. There are more silicon atoms in a grain of sand than there are grains of sand on a beach. Just for fun, prove this for yourself by picking some arbitrary number for grains per cubic inch and number of cubic inches in a good-sized beach.

If a grain is $1/50$ of an inch on a side, you can figure that there are $50 \times 50 \times 50$ grains in a cubic inch. Then you can figure the number of grains in a chunk of beach a half mile long by 50 yards wide by 20 feet deep. In inches, that's

$$(12 \times 5{,}280 \times {}^1\!/_2) \times (12 \times 3 \times 50) \times (12 \times 20)$$

That works out to about 1.7×10^{15} grains. One grain of sand is much less than a mole of silicon dioxide (I'm assuming plain old sand, not cool black sand from special lava or something like that). If you figure that it takes approximately 0.5 million of the little grains to make up a mole of silicon dioxide, then

$$\text{Silicon atoms in a grain} = \frac{(6.02 \times 10^{23})}{500{,}000} = 1.2 \times 10^{18}$$

So really, there are about a thousand times as many silicon atoms in a grain of sand than there are grains in a decent beach (I'm thinking specifically of Moonlight Beach in Encinitas, California, and would welcome funding for experimental testing of this proposition from July to September — you pick the year). In defense of beaches everywhere, I should point out that Copacabana is close to 5 miles long and almost 500 yards wide, making it a better numerical competitor with the single grain. But 20 feet is actually pretty deep for beach sand, so the grain still wins by a bit.

1/1,000

In the ideal world of Greek mathematics, numbers existed in a pure, theoretical realm. They could be calculated, in principle, to any degree of accuracy, but they existed as some perfect form in the eternal mind of the universe. Now that computers can do the arithmetic, performing all sorts of calculations with arbitrary precision is relatively easy. Want π to 40 decimal places? No problem at all.

The issue of precision is one point in which mathematics departs from the real world of *stuff*. In the real world, it's often difficult or expensive to get a measurement that's accurate to one part per thousand. Consider a few examples:

✔ **Rulers:** Head on down to a stationery store and find the flexible plastic rulers in the school supplies section. Pick out six of them and line them all up at the left-hand edge of the ruled part. Now look at the right-hand edge, where the rulers indicate 12 inches. Usually, there's a clearly visible discrepancy — after all, the process of making these rulers just involves printing on fairly soft plastic. Do you think that a plastic ruler can be accurate to one part per thousand (that is, one hundredth of an inch)? Heat one up in warm water and repeat the comparison if you do!

✔ **Lego® blocks:** If plastic objects have such sloppy dimensions, how can Lego make blocks that can be fitted into smooth walls hundreds of blocks high? Well, Lego needs to hold block tolerances of one part per ten thousand, and that means that Lego has to design its own plastic injection molds using its own special technology. As a result, the company produces toys that have greater precision than most plastic aerospace or medical parts.

✔ **Thermometers**: You used to be told that 98.6 degrees Fahrenheit was the absolute normal temperature. So that you could find out whether you were normal, classic mercury-based thermometers were developed to display only a few degrees over a measuring span of several inches. That made a tenth of a degree correspond to about a tenth of an inch, producing very high accuracy indeed.

Lately, measurements on tens of thousands of people have shown that the 98.6 number, faithfully and religiously repeated through decades in textbooks, simply isn't an absolute standard. Plenty of people have temperatures as low as 97 day in and day out, and plenty of other people hover between 99 and 100 degrees without having a fever.

✔ **Volume:** Special chemical laboratory glassware is calibrated to be accurate to about one part per ten thousand, as long as the volume is large enough. Kitchen glassware — a measuring cup, for example — is accurate to about one percent. These days, computer-controlled gasoline pumps report your gasoline purchase to a thousandth of a gallon, although a spot check by the state of California found the typical pump to be off by a tenth of a gallon or so. In other words, the pumps give you plenty of digits, half of which may as well be random.

The reason I think that $1/1,000$ is an important number is that it's the working definition of "good enough." If you know how much money you have to 0.1 percent accuracy, that's one part per thousand accuracy and should be good enough. One part per thousand means about one-fourteenth of an inch on a two-yard piece of cloth from a material store, and it's about the width of a line of grout on a tile floor in a standard-sized room. Keep this in mind when you look at results on a calculator — here in the real world, you can usually relate only the first three digits to physical stuff.

MCMLXXIV

It took an amazingly long time — hundreds of years — for the number system pioneered in India to work its way to Arabia and then across the Arab world to Spain, from which it was finally introduced to northern Europe. Contemporary reports suggest that the concept of zero was mind-boggling enough to give people headaches. These days we use "Arabic" numerals like 1, 2, and 3 without a second thought, and even kindergartners know about zero.

So except among mathematical hipsters, Roman numerals (which don't include a zero) persisted well past the year 1000 A.D. They still appear in movie credits, on medical school diplomas, and in several other areas where a little artificial augmentation of prestige is thought to be helpful. Decoding this particular number will at least help you with any number in this century.

You read it like so:

- ✔ M = 1,000.

- ✔ CM = 100 before 1,000, which means 1,000 – 100 in Roman. When a number is "out of order" (C is one hundred, and it's out of order because it appears before the M), you subtract it from the bigger number to its right. So CM is 900.

- ✔ L = 50.

- ✔ XX = 20. X is 10, XX is 20, and XXX is 30. If you had thought in terms of this very concrete kind of counting all your life, the notion of zero probably would have struck you as pretty spacey, too, just as it did the medieval monks who served as custodians of arithmetic.

- ✔ IV = 4. I is 1, II is 2, and III is 3. What more could you want? At 4, there's a little confusion. Sometimes it appears as IIII, the obvious choice, but sometimes it's IV, which is 4 according to the out-of-order rule since V is 5.

- ✔ The date MCMLXXIV is thus 1,000 + 900 + 50 + 20 + 4 = 1974.

The Romans, by the way, typically used an interesting design of abacus to do arithmetic, avoiding working out problems on paper. Using their humble abacuses and lots of common sense, they could solve most engineering problems. The number system wasn't very streamlined, but Roman engineers had a vast repertoire of tricks and work-arounds for calculations. As nearly as archaeologists can reconstruct the situation from Pompeii and from Rome's port city of Ostia, the Romans around 100 A.D. achieved a standard of living, at least in some cities, that was not matched again until around 1900.

$^9/_5$, 32, and $^5/_9$

Welcome to the U.S., the last place on earth to rationalize its measurements. Really, it's something of a miracle that we ended up with a decimal-based money system, approved in a short-lived fit of anti-British sentiment. I won't go on about inches and furlongs and quarts and pecks, but I must make some remarks about temperature. That a large chunk of elementary school science class has to be devoted to teaching the way scientists and the entire rest of the planet measure things is a pretty strange feature of American civilization.

We still use Fahrenheit temperatures, and everyone else (here called *EE*) uses the Celsius (also known as *centigrade*) scale. Both scales are based on the freezing and boiling points of water. On the Fahrenheit scale, 32° = freezing and 212° = boiling. Hey, what else?

On the Celsius scale, 0° = freezing and 100° = boiling, which in some sense certainly looks more promising. To get back and forth between the two scales, here's what you do.

Converting EE to U.S.

Their numbers are smaller. On a crummy, rainy spring day in Paris, the weather news may report a temperature of 5 degrees. The rule is: First you have to change the scale, and then you have to add a step.

Change the scale by multiplying the Paris number by $^9/_5$. That gets you

$$5 \times {}^9/_5 = 9 \text{ degrees}$$

Then you add the step, the 32-degree offset that Herr Fahrenheit built into his thermometers. That gets you to

$$9 + 32 = 41 \text{ degrees}$$

A rainy 41 degrees in Paris — sounds like Armagnac weather to me, kids.

By the way, if you don't feel like multiplying by $^9/_5$ in your head, just multiply by 2 instead. You end up with a number that's higher by a few degrees, but hey, it's just weather. Personally, I can't tell 41 from 43 degrees, anyway.

Converting U.S. to EE

The reverse process is this: Take off the 32-degree step and then change the scale.

Suppose you're trying to explain to an Italian why most of the American Southwest is uninhabitable without air-conditioning (some would say that it's *still* uninhabitable from a long-term ecological water-management viewpoint). You are making your point by discussing the week in Gila Bend, Arizona, where the temperature went past 122 degrees every day for a week.

First you step the number down by 32, so that 122 becomes

$$122 - 32 = 90$$

Then you change the scale. Remember that EE numbers are smaller than ours, so you multiply by $5/9$. That gives you 50 degrees. The last time your Italian friend saw that number on a weather map, it referred to some spot in Libya about 100 kilometers inland from the Mediterranean coast. Sounds like Campari and soda weather to him, assuming that he'd go near the place.

Again, if you don't like all this $5/9$, $9/5$ business, you can just divide by a factor of 2. In this example, you get 45 degrees Celsius. That still sounds like a blazing inferno to the EE community.

Megabytes

A whole new set of popular numbers has appeared in the wake of the computer revolution. Bits, bytes, and so forth are standard terms for measuring computer memory.

A *bit* is either a zero or one and can contain as much information as the state of an on/off switch — it's a one or a zero, so it represents on or off electronically.

A *byte* is a number composed of eight bits. Each bit can represent two states. That means that a byte can take any value (represent any number of states) between one and the number

$$2 \times 2 \times 2 \times 2 \times 2 \times 2 \times 2 \times 2 = 2^8 = 256$$

although, in computers, a byte is used to represent the numbers from 0 to 255.

That's a pretty good range, because it can represent all the characters and numbers typically used in printed English. English and European languages, in the world of hard-core computer designers, are called *single-byte languages* because they can be represented with a single-byte alphabet. Because they have much larger character sets, Japanese and Chinese are called *double-byte languages* because two bytes can represent 65,536 characters. That's more than enough to look up the 40,000 characters used in Chinese and Japanese books in a computer-based lookup table. So you can assign a two-byte number to each character in a Japanese dictionary and thus represent Japanese (with some difficulty) inside a computer.

A megabyte is approximately a million bytes. More exactly, it equals

$$2^{20} = 1,048,576$$

Back in the late 1980s, a computer with 2MB of memory and a 200MB hard drive was considered pretty fancy. Now I'm constantly asked by people at my computer-magazine job if such a computer is even good enough to do elementary school homework. It's a very odd business, really — the computer industry has convinced everyone that computers that would have been the Big Computers at NASA a decade ago aren't fit for kindergartners. This attitude has more to do with marketing than reality.

666

This is the number of the Beast in the Book of Revelations, and it's a great starting place for mystical speculations. The game here is to decide who the Beast is by assigning numbers to letters in a name and then adding up the letters to get 666. It is widely believed that the first version of this was meant to refer to the full Latin name of the emperor Nero.

By the time of the Reformation, Catholic scholars had determined that one spelling of the name of Martin Luther worked out to 666, and Protestant scholars returned the favor by finding 666 in the name of Pope Leo X. As you might expect, numerologists have associated the number with Adolf Hitler, Franklin Roosevelt, and a host of less likely Beast celebrities, on down to Rush Limbaugh and Microsoft's Bill Gates.

One conventional assignment for this kind of calculation is $a = 1$, $b = 2$, up to $z = 26$. Sometimes you have to assign $a = 26$ and so forth instead of $a = 1$ to make things work, and sometimes you have to cross out the letters that aren't used in Roman numerals, but of course it's well worth the effort.

A similar speculation concerns dates for the end of the world, based on internal Biblical clues. For current apocalyptic fans of the year 2000 as the end of the line, I would like to suggest that our calendar has almost certainly misassigned the year zero by at least five years one way or another, so either it's already ended and no one told us or we have a bit more time left.

1,729

You owe it to yourself to read Robert Kanigel's wonderful book *The Man Who Knew Infinity* (Scribner's, 1991), a biography of the Indian mathematician Srinivasa Ramanujan. You may be thinking to yourself, "Actually, I seldom read biographies of Indian mathematicians." This book is different.

Trust me; you'll like it a lot. Ramanujan was almost self-taught and discovered a large chunk of modern mathematics by himself, including many topics of current interest in computer design and physics. At one point in his unbelievably brilliant career, he ended up in Cambridge, England, where neither the food nor the weather nor the outbreak of World War I did him much good.

While in a hospital in England, Ramanujan was visited by his coworker G. H. Hardy, one of the leading mathematicians in England in the first part of this century. Hardy remarked that he had arrived in taxi cab number 1,729, said that it seemed like an uninteresting number, and hoped that it wasn't a bad omen. Ramanujan said, "No, it's a very interesting number. It's the smallest number that can be expressed as the sum of cubes in two different ways." That is, he had figured out in a fraction of a second that

$$1,729 = 1^3 + 12^3 = 10^3 + 9^3$$

The trick was not just to have seen this but also to have figured that there are also no numbers smaller than 1,729 for which you can make a similar decomposition. And hey, the guy was lying there with the flu! I should also emphasize that Ramanujan's contributions were all serious, almost weirdly original mathematics in infinite series, continued fractions, and related topics.

The Golden Ratio

Greek mathematicians concluded that a rectangle in which the long side is 1.618 times longer than the short side is particularly aesthetically appealing. This ratio appears in the construction of Greek temples, not just in the overall proportions but in the shapes of doorways and windows. Although it's not clear that the Egyptians had the same mystical associations with this number that fascinated the Greeks, the number also appears in pyramid construction. A large-scale 19th century study of all sorts of rectangular shapes (bricks, playing cards, greeting cards, and so forth) from cultures all over the world concluded that people have an intuitive preference for this kind of rectangle compared to rectangles that are either shorter or thinner. Shockingly enough, *legal-size* paper is 1.64 times as long as it is wide — that's close enough!

The exact number derived by the Greeks is

$$\Phi = \frac{(\sqrt{5} + 1)}{2}$$

$$= 1.61803...$$

So 1.618 is certainly a close enough approximation if you want to make your own Greek temple. Figure 23-1 shows how this number is used to construct a rectangle and spiral.

side length = 1

side length = 1.618

The Golden
Rectangle:
Classical
Proportions

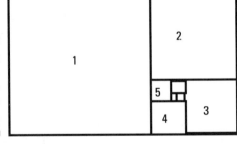

10^{14}

Just a year ago, the astrophysics community generally agreed that the universe was 10^{14} years old. Then things got strange. First, using standard techniques but new information from space-based telescopes, astronomers found objects that seemed to be older than the "established" age of the universe. And then they observed objects that appeared to be farther away than any yet observed but that were ridiculously young by astrophysical standards. According to the pretty-much accepted Big Bang theory, you are supposed to see the old stuff, not new stars, at the outer edges of the universe.

Reasonable scientific evidence (setting aside for the moment possible religious objections) shows that the Earth is about 4×10^9 years old, so it's only been around for a blip in terms of the history of the universe. It's the rest of the sideshow where the dating is more or less out of control at this point. I put this number in the collection because, unlike π or 666 or 1,729, you'll see it toggled up and down in newspaper accounts of the latest theories in cosmology. Please feel free to update it every six months or so in your own copy of this book.

Chapter 24

Ten Calculator Tips

*T*here are a great variety of calculators on the market, from simple four-function calculators ($+, -, \times, \div$) that are given away free with a pack of ball-point pens to programmable scientific calculators that compete with small personal computers.

All calculators are *not* created equal. You may be shocked to know that they are not all equally accurate, but you may be pleasantly surprised to find that you can mimic many of the functions of fancier calculators on the cheapest four-function units available. Anyway, this chapter provides some considerations, tricks, gimmicks, and serious points for calculator users.

How Good Is Good Enough?

At the low end of the calculator food chain are lots of calculators that are accurate to eight digits. When you consider that these calculators are supposed to be used mostly as adding machines, this isn't a problem — you can keep track of dollars up to $99 million without a mistake. And, of course, once you get to the $100 million mark, you can afford to go out and spend $12 for a modest financial calculator, one that has keys for PMT (payment) and so forth. What the heck, live a little!

In the meanwhile, here's how to check out the numerical accuracy of the calculator you're using.

Divide the number 1 by the number 666 (not picked because it's beastly — it's just a good demonstration). You will probably see the following:

$$1 \div 666 = 0.0015015$$

Now, if you multiply this result by 666, you should get back to the number 1, right? More likely, you'll come up with this:

$$0.0015015 \times 666 = 0.999999$$

The reason is that the calculator would have had to carry at least two more digits (digits that don't appear in the eight-place display) to make a number that the calculator would "know" to display as 1.

You may think that this small discrepancy isn't so bad. In that case, try another example: 1 divided by 1,666.

$$1 \div 1,666 = 0.0006002$$

Now, when you multiply this last number by 1,666, you come up with this:

$$0.0006002 \times 1,666 = 0.9999332$$

Clearly, this answer is not the same thing as 1. The problem here is that to get this number to round off to 1 in the calculator display, the calculator would have to keep three more decimal places than it shows in the display (it would keep the extra digits in memory). It rounds off the answer to the eight digits in the display, and that accuracy isn't good enough to recover the original number.

What I said earlier about using calculators as adding machines still holds — you're not going to get addition errors. But for long series of multiplications, this sort of small error tends to accumulate. Every time you do a repeated division or multiplication, you get a little bit of error-creep, which in practice can easily make it into the third place after the decimal. For figuring mortgage payments to the penny, for example, you're better off buying a real financial calculator with keys for interest rates and payments than using a four-function calculator and plugging numbers into complicated formulas.

More on Accuracy

A scientific calculator, one with keys for sine, cosine, logarithm, and so forth (you probably had to buy one in high school), is usually more accurate than a four-function calculator. How many more decimal places beyond the digits in the display does a scientific calculator keep? Here's one way to find out. Call up π (pi) — nearly all scientific calculators have a key for π.

Try the sequence

$$\pi = 3.1415927 - 3.1415926536 =$$

Sometimes you get the result zero. Sometimes you get the result

6.0E–10

From scientific notation (the display with E in it), this translates to 0.0000000006. This means that despite the displayed number 3.1415927, the calculator really keeps the internal value 3.141592654. This is also true for many other "mystery" numbers, with the prime example being

$e = 2.7182818$

This number turns up over and over again in compound interest and other calculations based on growth formulas.

Science Students, Beware: Do Your Answers Make Sense?

This book may have fallen into your hands while you're still a student. Teaching is not its main aim — its main purpose is to try to restore a bit of math to people whose only memory is, "At school, they proved to me mathematically that I'm hopeless."

Nonetheless, I'm slipping this part in as a public service to my brave former colleagues laboring away teaching science in high schools and universities.

Here's the point:

Whenever you use your calculator, stop and think if the answer you are giving could actually

- ✔ Be possible
- ✔ Be measured

Is it possible?

I have seen students hand in tests in which they computed amounts of heat released in a chemical reaction that exceeded the heat of an H-bomb. In physics classes, it is not unheard of for students to find stones dropping from a building and subsequently exceeding the speed of sound on impact. This is the science class equivalent of calculating the mortgage payments on a $120,000 house and finding payments of either $2.17 or $53,789 per month.

When you see a problem on a test or even in a homework assignment that calls for a numerical answer, test your understanding of the problem by seeing whether you can write down what would be a reasonable range for the answer. A rock thrown from a tall building, for example, is going to hit the ground with a speed somewhere between 40 and a few hundred miles per hour (it's your job as a student to know what these numbers are in meters per second, too).

No matter what you find on your calculator, 0.24 miles per hour isn't a plausible answer, and neither is 24,000 miles per hour. If you get this point straight, you can have a great time on standardized tests like the SAT because the test designers usually include two ridiculous answers in each set of five. Immediately ruling out the ridiculous answers virtually guarantees that you will do well on such tests.

Is it measurable?

Another calculator problem that students encounter is reporting too many decimal places. Try this very basic physics problem:

A snail is in a desperate hurry to cross a driveway in the rain before a cat spots him. Therefore, he is hustling along at uncommon speed, covering 10 centimeters (cm) in 3 seconds. What is the snail's speed?

If you report the answer as 10/3 = 3.33333333 cm/sec, you have made an error that's almost as bad a reporting the snail's speed as 30 miles per hour. The data in the problem has only two decimal places, so you are justified in reporting the answer only as 3.3 cm/sec, no matter what the display on your calculator says. In fact, it would be a challenge to a well-equipped physics laboratory in a university to measure a speed like this to six decimal places, much less eight or nine.

The chemical equivalent problem asks for the molar concentration of a solution of 32 grams of sodium hydroxide in 500 milliliters of water. After working out the molecular weight of sodium hydroxide correctly (it's 40), it's pretty common for a student to report a concentration as

C = 1.66666666 M (moles per liter)

Technically, this answer is more ridiculous than the one in the physics case. In real student labs, the displays on lab balances report only four digits, and the chemicals themselves pick up enough water from the atmosphere between the chemical jar and the balance to make the final weight of the chemical uncertain in the third digit of the reported weight displayed by the balance.

An extremely careful weighing operation with an electrobalance inside a zero-humidity dry box will give five accurate decimal places for the weight of a chemical. Getting accurate measurement of liquid is an even worse problem. Using existing technology on planet earth, there is no way to make a 1.66666666 molar solution of anything. When you get points marked off on a report or test for putting down an answer like this, you deserve the hit.

Square Roots

Some plain four-function (+, −, ×, and ÷) calculators also have a $\sqrt{}$ key, but most of them don't. I just want to step through the guess-and-divide method for finding square roots because this is one of the few functions that turns up a lot in daily life. Every time you cut a piece of plywood on a diagonal or figure how much roofing material you'll need, a square root is lurking in the underbrush.

The rule

Here's the method: Look at the number for which you need a square root, say 8. Guess any number for a root (this first guess doesn't have to be good) — I could try 3. Divide it into the original number, giving 8/3 = 2.66. Take the answer from this operation and average it with the earlier number (the average of 3 and 2.66 is 2.83). Repeat the process by taking this new average (2.83) as the starting point.

An example

For the sake of concreteness (or maybe I should say woodenness), I'm going to find the corner-to-corner diagonal length of a standard four-foot-by-eight-foot sheet of plywood.

One side of the sheet is

$$4 \text{ feet} = 4 \times 12 = 48 \text{ inches}$$

and the other side is

$$8 \text{ feet} = 8 \times 12 = 96 \text{ inches}$$

Looking at the picture in Figure 24-1 and invoking the Pythagorean theorem ($c^2 = a^2 + b^2$), I have

$$\text{Diagonal}^2 = 48^2 + 96^2$$

$$\text{Diagonal}^2 = 2{,}304 + 9{,}216$$

$$= 11{,}520 \text{ inches}^2$$

Now, I don't want inches *squared* — I want to know how long the diagonal is in inches, so I have to find the square root of 11,520. Also, because I'm just trying to get numbers in inches, I keep only one place after the decimal. If you've ever used a SkilSaw, you should be aware that tenths of an inch are a bit of a challenge to replicate in wood with construction-grade tools.

8 ft

4 ft

Figure 24-1:
Plywood
and the
Pythagorean
theorem.

Plywood and Pythagoras: the diagonal
length is the square root of 48^2
inches plus 96^2 inches

To find the square root of a number such as 11,520, I just go through the steps that I just outlined, repeated here:

1. **Make a guess.**

   I'm going to guess 100, just because one side of the triangle was 96. That means that 100 has to be somewhere in the neighborhood.

2. **Divide the guess into the number you're trying to find the square root of (in this example, 11,520).**

   $$11,520 \div 100 = 115.2$$

   Actually, you could have done this without a calculator! Dividing by 100 just means moving the decimal point over two places.

3. **Average your first guess and the number that you find by dividing that guess into the original number.**

   In this example, my guess was 100 and I found the average to be 115.2, so

   $$\frac{(100 + 115.2)}{2} = 107.6$$

   That means that 107.6 is the new guess.

4. **Repeat the process.**

   That means to divide 107.6 into 11,520:

   $$11,520/107.6 = 107.1$$

Actually, the calculator says 107.06319, but I rounded it off to 107.1 because 0.06319 inches of plywood doesn't mean a lot to the average construction craftsperson.

The process "converged," as the mathematicians say, in only a few steps. That means that it produced a usable answer, and it produced it very quickly.

Food, Glorious Food

You probably can't believe that I'm going to bring up dieting after all this noise about square roots, but I am. This section is in here by request — my jolly study group in Sonoma County insisted that I include it.

Your body (he began portentously) is like a checkbook. That line would make a great introduction to a high school health movie, wouldn't it? You make deposits (food), you make withdrawals (activity), and you pay a sort of constant daily checking fee (resting metabolism).

To stay at a particular body weight, you need a certain number of calories from food, assuming a base of almost no physical activity. If you take in more calories than this "maintenance level," you start gently drifting up in weight until you wake up one morning in your early 40s about 30 pounds heavier than you were in college.

Your choices at that point are pretty basic: Take in less food or engage in more activity. Fairly active exercise — walking along modestly quickly — burns about 200 calories per hour. Really killing yourself is good for 500 to 600 calories. If you take in 3,500 calories less than maintenance level (or burn the equivalent in extra activity), you lose a pound. Here's the calculator problem:

Three cups of Haagen-Dazs butter pecan ice cream has 1,896 calories. According to the digital readout on the fabulous new exercise bicycle you just bought, setting 3 in "uphillness" is worth 580 calories per hour. How many hours do you have to move your sweating, sorry carcass on this fiendish device to work off the one miserable cup of ice cream?

It's a simple division problem:

hours = 1,896 ÷ 580 = 3.27 hours

Just to be on the safe side, may as well make it four hours. Hey, watch the movie *Gettysburg* while you pedal.

The same way you shouldn't make minimum payments on large credit card balances at 21 percent, you shouldn't eat real ice cream if you want to lose weight. Ice cream not a problem? A six-pack of Michelob is 960 calories. It's easier not to consume it than to work it off. It's just math.

By the way, one of the really annoying problems in human physiology is that your body adapts a bit to diets. Stop eating, and your calorie requirements drop somewhat. But they don't drop by a huge amount because you still need calories to maintain your body temperature and to operate your nervous system. The remarkable fact about your nervous system is that solving differential equations takes no more calories than watching *Ren and Stimpy*.

The Calculator Alphabet

The seven-segment characters in a calculator display lend themselves to an upside-down alphabet interpretation. The correspondence is as follows:

- B = 8
- E = 3

- ✔ G = 6
- ✔ H = 4
- ✔ I = 1
- ✔ L = 7
- ✔ O = 0
- ✔ S = 5
- ✔ Z = 2

Over the years, people have made up all sort of entertaining and goofy puzzles using this letter code. Would I 1585 ÷ 5 to you?

Adapting Recipes for More or Fewer People

If you really, really have a hard time converting recipes — say a recipe for four people that you want to make into a recipe for six — you have two choices. As an American, you face a rich birthright of funny business in recipes, from teaspoons of this to $5^1/_4$ cups of that.

Choice one

Just give up on fractions. If the recipe says $5^1/_4$ cups, convert it to 5.25 cups. To scale this up to 6 people from 4, you just multiply as follows:

quantity = 5.25 × (new amount/old amount) = $5.25 \times {}^6/_4$

= 7.875

You are unlikely to have measuring cups marked in decimal fractions, so you have to use a bit of sense. For example, 7.875 cups is just a bit under 8 cups (okay, it's $7^7/_8$ cups). So just use a bit less than 8 cups. Any recipe worth using can survive at least 10 percent imprecision.

Choice two

For about ten dollars, you can get a calculator that does fractions directly. Look for a key that says

a b/c

You can then enter fractions like $3^2/_7$ directly and multiply them by $^{21}/_5$ if you like. The calculator reports the result in any fractional form you want. Problem solved!

Finding Sines

With a humble four-function calculator, you can do a pretty good job of approximating trigonometric functions.

Look at this short list with the angles in degrees:

- sin 0 = 0.000
- sin 15 = 0.259
- sin 30 = 0.5000
- sin 45 = 0.707
- sin 60 = 0.866
- sin 75 = 0.966
- sin 90 = 1.000

Even this short list is sufficient to let you interpolate sines of other angles. Suppose you needed the sine of 7.5 degrees. You can look in this list, find 0 for 0 degrees, 0.259 for 15 degrees, and guess that the sine of 7.5 would be halfway between. You're right, too. Getting sines correct to eight decimal places takes some ingenuity on the part of calculator manufacturers, but getting a "good enough" number to two decimal places is simple. I have never understood why some enterprising protractor manufacturer didn't simply print the sines and cosines of angles right there on the plastic — it would make the protractor useful in both geometry and trigonometry classes.

Numerical Silliness

Have someone write down a three-digit number, say 714, and then repeat it, writing 714,714.

Then have that person enter the number on a calculator, and explain that, as a result of mastering the contents of *Everyday Math For Dummies,* you can now identify factors of numbers (factors divide evenly into the number in question), among other feats of numerical prowess.

Stare at the number for a few seconds and then say, "Okay, right off the bat I can see that 7, 11, and 13 are factors of that number, and by the way, those numbers (7, 11, and 13) are all primes." The person with the calculator can confirm that these factors divide evenly into 714,714. Amazing!

What's going on here is that writing the number down in this way amounts to multiplying it by 1,001; and 7, 11, and 13 are the factors that make up 1,001.

Reverse Polish Notation

Calculators at the low end of the market follow the ordinary algebraic style of calculation entry. If you want to multiply 23 by 15, you enter 23, then press the × key, then enter 15, and then press the = key.

Hewlett-Packard, a long-time calculator pioneer, made and still makes calculators with a different entry style called *Reverse Polish Notation* (RPN). The name pays tribute to the work of a Polish logician named Lukasiewicz, but in fact this mode of entry is the direct descendant of the workings of old Friden mechanical calculators.

The keys for the preceding problem would be

23

[Enter]

15

×

in that order. You save a keystroke by doing the problem this way, and a lot of Hewlett-Packard's motivation in choosing RPN is that using RPN on programmable calculators enables you to pack fancier programs into a much smaller space. I bring up this topic just as a piece of calculator advice. HP calculators are nearly bullet proof (I have one from 1977 that still works!) and have tons of features, but the RPN models require that you make a small adaptation to your calculating methods.

"IT SAYS HERE IF I SUBSCRIBE TO THIS MAGAZINE, THEY'LL SEND ME A FREE DESK-TOP CALCULATOR. DESKTOP CALCULATOR?!! WHOOAA—WHERE HAVE I BEEN?!!"

Chapter 25
Ten Topics in Advanced Math

*W*hen you're finished with this book, I expect you to know why you don't want to leave a huge balance on a 21 percent interest credit card. I also figure that you should be keeping your checkbook in order and are calculating your own tips.

This chapter, on the other hand, is more of a guided tour of topics that are really the property of someone else. It can't hurt to have heard of some of these things, a few of which are stuffy old subjects from the traditional curriculum and some of which are the latest rage in computer math.

Complex Numbers

Officially, I could have made some remarks about complex numbers either in the algebra or the trigonometry chapter since you probably heard about them in high school. I didn't for the fairly straightforward reason that, although they're important, you don't see much of them in daily life unless you're an electrical engineer.

So here's the story. You already know something about square roots. You know that

$$\sqrt{4} = 2$$

and with a calculator you can show that

$$\sqrt{10} = 3.1623$$

Imaginary numbers provide a definition for the square root of a negative number. The number *i* is defined as

$$\sqrt{-1} = i$$

To show the kind of distinctions that I have spared you of thus far, I'll tell you that the preferred format of this definition is

$$i^2 = -1$$

Actually, it's considered better to say that the square root of –1 isn't really defined but that the square of *i* is. Similarly, the quantity $^1/_0$ is officially undefined, but in loose practical terms it's infinity (that's what your calculator will say, or overflow). Infinity isn't really defined either, but as you divide 1 by smaller and smaller numbers, the result gets bigger and bigger.

One of the main applications of imaginary numbers is in defining a new number plane that has an astonishing range of properties and applications. Look at Figure 25-1. The *y*-direction in the usual Cartesian plane in geometry or graphing has been relabeled the *iy*-direction. The point (3,4), which used to stand for *x* = 3, *y* = 4, now indicates *x* = 3, *iy* = 4. All the numbers in this plane can be represented in the form

$$a + bi$$

Multiplying imaginary numbers in this form is not particularly hard. Look at this example:

$$(2 + i)(3 + 2.5i)$$

$$= 2 \times 3 + i \times 3 + 2 \times 2.5i + 2.5 \times i \times i$$

$$= 6 + 3i + 5i + 2.5\,(i^2)$$

$$= 6 + 8i + 2.5\,(-1)$$

$$= 3.5 + 8i$$

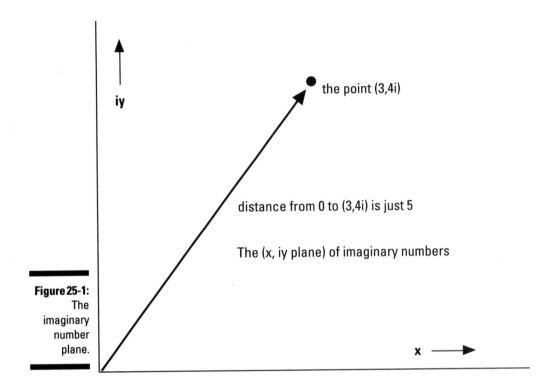

iy

the point (3,4i)

distance from 0 to (3,4i) is just 5

The (x, iy plane) of imaginary numbers

x

Figure 25-1:
The
imaginary
number
plane.

When complex numbers are expressed in another form, involving distance in the plane from zero and angle from the *x*-axis, it's even easier to multiply them. That, however, is another topic pretty far from everyday math.

Types of Infinity

Just in the preceding section, I said that infinity isn't really defined. That's not exactly true, because many great mathematical minds have spent lots of time trying to get a grip on this concept. One school of modern mathematics, called *constructivism,* explicitly rejects the concept of infinity. In the constructive approach, if you can't show how you come to a conclusion in a definite number of steps, then you don't really have a conclusion. This approach admittedly throws out a lot of mathematics, but it fits the realities of mathematics done on a computer, where infinite processes take infinite time and thus aren't very valuable.

Here are a few points to ponder. Consider the whole numbers

$$1, 2, 3, 4, 5, 6, \ldots n$$

There isn't any "biggest number" when you approach the problem this way. At any point, you can add one to the biggest number you have, and you have the next bigger number. The whole numbers are thus described as an infinite set because you can't run out of them.

Now, are there more fractions than whole numbers? If you think about it for a minute, you can see that you can put the numbers

$$1^1/_3, 1^2/_3$$

between the one and the two. So are there more fractions, specifically thirds, given that you can fit two of them after every single whole number? Are there twice as many of them?

In the late 19th century, the mathematician Georg Cantor formulated the argument shown in the diagram of Figure 25-2. This argument suggests that there are as many fractions as there are whole numbers because you can match them up one to one. The set of whole numbers and the set of fractions are both infinite, but they are said to be of the same order of infinity because of this correspondence.

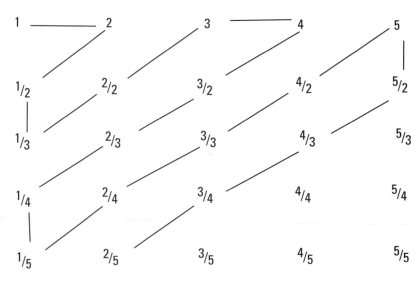

Figure 25-2:
An infinity of whole numbers and fractions.

Cantor further showed that you can't make this kind of correspondence between the real numbers (the set of fractions plus the irrational numbers like $\sqrt{2}$ and π). He argued that the interval between zero and one contains infinitely more numbers than the set of whole numbers and went on to characterize still-higher orders of infinity. Just as the real numbers are somehow "more infinite" than the whole numbers, it's possible to construct "more infinite" sets than the real numbers.

You're about as far away from the constructive approach as you can get with these arguments, and, to a certain extent, constructivism grew out of a reaction to the paradoxes that result from indirect reasoning about infinity. The constructivists decided that it would be very difficult in some cases to be sure that arguments about different kinds of infinity would be correct, so they decided, "Let's not do this." Meaning no disrespect whatsoever to a great mathematician, but many people who grappled with Cantor's original papers noted with alarm that he ended his days confined to a psychiatric hospital.

What Calculus Does

You may not have gotten to calculus in high school, but if it's any consolation to you, calculus is easier than it looks. This section provides just a museum tour so that you'll know what it looks like when you see it. You probably don't know a lot about rhinoceroses, but you can tell them from hippopotamuses. And even if you don't know much calculus, you'll be able to tell that most of the math in "Far Side" cartoons is in fact gibberish. I'll never understand why cartoonists don't copy real formulas from textbooks.

Integral calculus

At the lowest level in applications to physics and engineering, integral calculus turns out to be a set of formulas that compute areas, volumes, and sums of series. One of the simplest cases is the area under the line $y = x$ (see Figure 25-3). One argument is based on dividing that area into little rectangles of known height and width and then summing up all these little rectangular areas to a particular value of x (see Figure 25-4). This argument gives

$$Area = \frac{x^2}{2}$$

as the area up to any given value of x. You can see for yourself that the formula is right by noticing that the area under the line $y = x$ is always a right triangle of height x and base x also. Since it's just a triangle, the formula for the area of a triangle applies and is just

$$\text{Area} = \text{base} \times \text{height}/2 = (x)(x)/2$$

The plot and formula in Figure 25-5 summarize all this information.

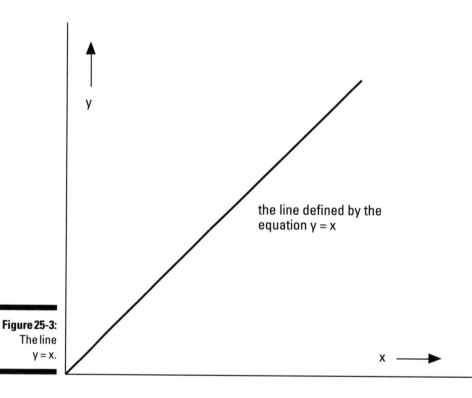

the line defined by the equation y = x

Figure 25-3:
The line
y = x.

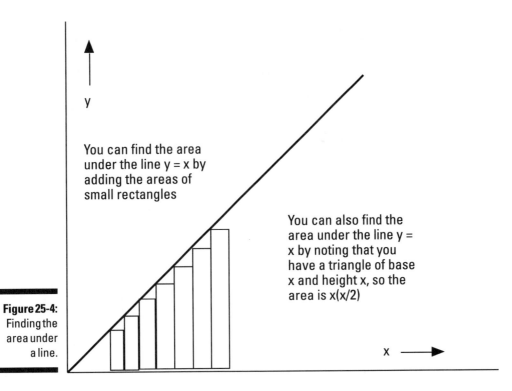

You can find the area under the line y = x by adding the areas of small rectangles

You can also find the area under the line y = x by noting that you have a triangle of base x and height x, so the area is x(x/2)

y

x

Figure 25-4:
Finding the area under a line.

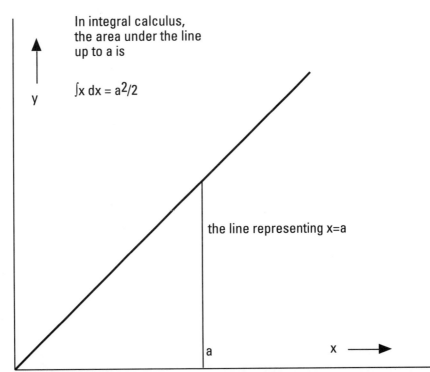

In integral calculus, the area under the line up to a is

$$\int x \, dx = a^2/2$$

y

the line representing x=a

a

x

Figure 25-5:
An integral and a formula.

Engineers use integral calculus to determine the volume of cones, cylinders, spheres, and egg-shaped containers, although the formulas for common volumetric objects are already known and no one actually has to solve calculus problems to determine these formulas. You can tell integral calculus from yards away simply by looking for the long, heavy S — the integral symbol.

Differential calculus

Differential calculus concerns rates. One of the simplest examples involves the speed and distance that a rock dropped from a tower travels. (This is the basic problem that Isaac Newton addressed.) Look at Figure 25-6, in which the problem and terminology are expressed in traditional notation.

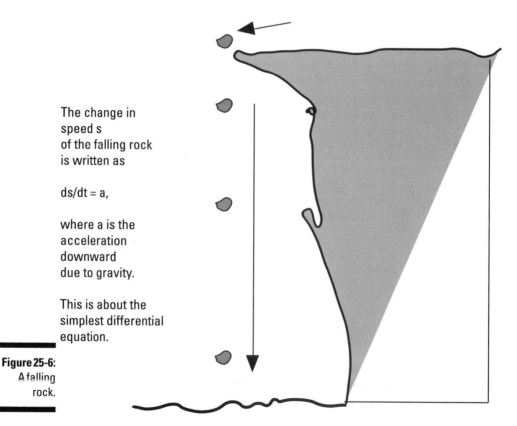

The change in
speed s
of the falling rock
is written as

ds/dt = a,

where a is the
acceleration
downward
due to gravity.

This is about the
simplest differential
equation.

Figure 25-6:
A falling
rock.

According to the results determined in this figure, the rock just keeps picking up speed. In real life, the resistance of the air gives the rock a *terminal velocity,* a speed that remains constant for the rest of the drop. A feather has a low terminal velocity, a rock has a high one, and you personally, should you jump off a tower, have a terminal velocity that's in between these two values (you're less dense than a rock).

A *differential equation* is just an expression that defines a relation between rates. The standard method of solution in the old days (again, of course, all the simplest equations have been solved) was to guess at an expression that would satisfy the relation. In the computer age, the usual method is to set out some starting conditions and then take the equation as a programming instruction, literally following the changes that are specified in the differential equations step by step (see Figure 25-7). That some differential equations are amazingly badly behaved has turned out to be one of the big surprises of the last 20 years — little changes in starting conditions produce big changes very rapidly in the solution curves.

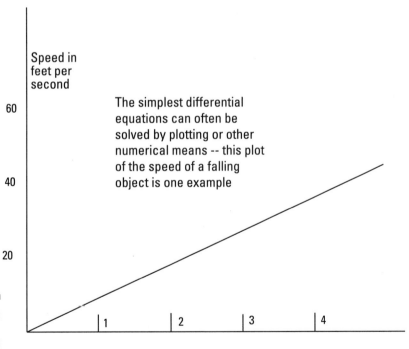

Speed in feet per second

The simplest differential equations can often be solved by plotting or other numerical means -- this plot of the speed of a falling object is one example

Figure 25-7:
Computing
differential
equations.

time in seconds

Symbolic Math Programs

You already know that if you need to find the square root of 17, you can do so with a pocket calculator. You may also expect that if you need to find the optimal shape for a tanker truck — you have probably noticed that they often have interesting mathematical shapes — you can find the shape with some type of computer program.

The new motto is "Computers, they're not just for numbers anymore." Not only can you ask a computer program to expand an expression like

$$(a + 5)^9 =$$

but you can ask such a program to find a solution to the equation

$$ax^3 + bx + c = 0$$

in terms of the unspecified constants a, b, and c.

Figures 25-8, 25-9, and 25-10 just show off some of the capabilities of symbolic mathematical programs, respectively Mathematica, Theorist, and Maple. I picked these because they run on every common personal computer and because each one is available in an inexpensive student version. It's a safe bet that even if you were taking mathematics courses in graduate school, these programs could take on every math problem you would ever encounter.

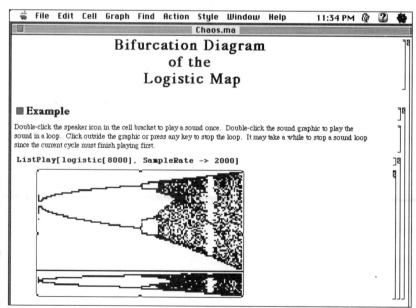

Figure 25-8: Mathematica.

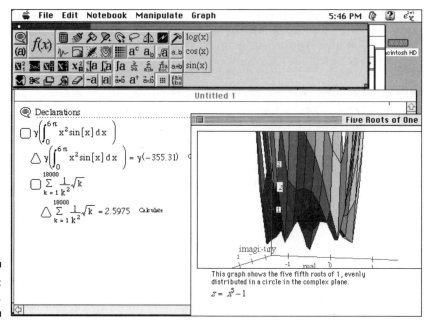

Figure 25-9:
Theorist.

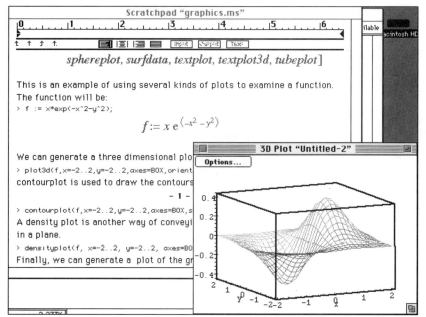

Figure 25-10:
Maple.

Curiously, even though these programs can do tricks that the great mathematicians of the past scarcely dreamed of, only a handful of people have used them to produce original, new mathematics. The obvious conclusion is that real mathematical talent is fairly rare. Somehow, just as the typewriter didn't produce a revolution in poetry and word processors do not seem to be upgrading the quality of modern novels, math programs seem not to be producing a Renaissance in pure mathematics, but they're doing great things in applied math and engineering.

A Bit of Statistics

Tragically, statistics is regarded as a difficult subject. You gain almost enough ordinary experience in life to guess a great number of the most important statistical results, but you are almost certain to be put off by much statistics terminology.

Normal IQs and such

The curve in Figure 25-11 is called a *Gaussian curve,* or a *normal distribution,* and is sometimes informally called a *bell curve.* (For a review of the notorious 1994 book called *The Bell Curve,* I recommend, as a professional statistician, the book review in the February 1995 issue of *Scientific American.* For a short review, I can tell you that the book is a rehash of discredited political arguments dating back to the Civil War and contains nothing that a statistician would consider to be appropriate treatment of data, much less actual scientific evidence.)

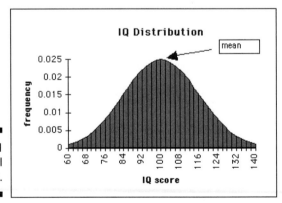

Figure 25-11
The normal distribution.

This curve is characterized by a *mean* value (refer to the figure), which is the average point in the distribution. It's also characterized by a *standard deviation,* which is a measure of the width of the distribution. To take a concrete example, the normal distribution that fits standard IQ tests has a mean value of 100 (loosely, the average IQ is 100) and a standard deviation of 16 points. Because the curve has a mathematically defined shape, it's possible to find the area under it for different values of x by using the formula for the curve and integral calculus.

That, in turn, means that it's possible to determine the fraction of the population with IQs above some fixed number or below some fixed number. The organization Mensa, which tries to select members with high IQs, accepts test results that are two standard deviations above the mean, meaning 2 times 16 (32 points) above the mean of 100, or IQs above 132. It happens that about 2 percent of the population has IQs of 132 and above, implying that 98 percent of everybody has an IQ lower than 132. This definition (the top 2 percent) is usually the definition that high school counselors use for applying the term *gifted,* although nobody in the cognitive sciences would agree with that definition any more.

Normal piston rings and quality

This IQ business just happens to be one everyday use of the normal distribution. Another common use is in quality control. Suppose, for example, that you are manufacturing piston rings. If you carefully measure the diameter of a large set of piston rings, you will find a bit of manufacturing variation, as shown in Figure 25-12.

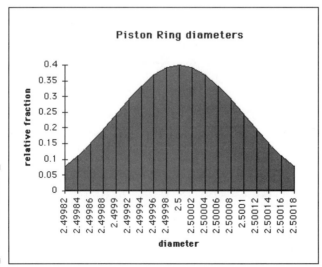

Figure 25-12: The normal distribution of piston rings.

Here, the mean is 2.5 inches and the standard deviation is 0.0001 inch, or one ten-thousandth of an inch. With better manufacturing methods, you can aim to reduce this standard deviation, making the parts more strictly uniform. In fact, that's pretty much the philosophy of the industrial quality control methods perfected by the American statistician W. Edwards Deming.

Deming proposed that with tighter statistical specs, it would be easier to produce superior designs. He had a great influence on methods of production during World War II. After the war, in the Fabulous Fifties, American markets displayed a nearly limitless appetite for goods, and Deming's consulting services fell out of favor in the U.S. So he went to Japan and managed programs that turned Japan's output from a synonym for junk to the most rigorously quality-controlled manufactured goods in the world. The annual Japanese quality-control prize is called the Deming Award.

At one point in the 1970s, Ford Motor Co. bought a stake in the Japanese company Toyo Kogyo, which was supplying the small truck sold in the U.S. as the Ford Courier. A team from Ford visited Toyo Kogyo several times and finally asked, out of plain curiosity, who was the white-haired old gentleman (Deming looked a bit like Colonel Sanders of Kentucky Fried Chicken fame) in the pictures in everyone's offices at the Toyo Kogyo works. The Japanese executives couldn't believe that the Ford executives didn't recognize the famous American quality expert Deming. They may have believed it if they'd had an opportunity to drive my very own 1978 Ford Mustang II. From door handles to paint job to U-joints to electrical system, it was a veritable rolling laboratory in (lack of) quality control. Things are better now at Ford than they were in the '70s, I'm thrilled to report.

Game Theory

Fortunately, there's a great book on this topic that has plenty of everyday applications. The book is partly a biography of the brilliant 20th-century Hungarian mathematician John von Neumann and partly an explanation of game theory.

The book is called *Prisoner's Dilemma* by William Poundstone (Doubleday, 1992), after the first basic strategy game analyzed by von Neumann. A picture of the choices is shown in Figure 25-13. The situation is this: The police have hauled in two suspects in a crime. If both suspects hold out and refuse to confess, they might be released. If suspect A rats on suspect B, suspect B may go to jail. If B rats on A, it's the reverse. If they both rat on each other, it's even worse for both of them. The problem is that each suspect has to make a decision without knowing what the other did. In the best case, they hold out and refuse to confess, but there's the temptation to go for a smaller penalty by cooperating. What to do? Hey, that's why it's called a dilemma.

Prisoner's Dilemma

The police have arrested two burglars they suspect of working together. Burglar A and Burglar B are put in separate rooms for questioning. How do they decide whether to confess and rat out their partners, or stonewall? Here is the result table.

	B confesses	B doesn't
A confesses	They both get ten years.	A serves no time, B gets 5 years.
A doesn't	B serves no time, A gets 5 years.	They both get one year.

Figure 25-13: A bit of game theory.

As it happens, a huge number of transactions in real life can be modeled on a set of boxes like those in Figure 25-13. Many optimal results occur with cooperation, but there are lots of cases in which you try to protect yourself from cheating on the part of a second party. von Neumann's game theory is the basis not only of the most recent round of Nobel Prizes (to one other Hungarian, too) in economics, but it underlies most of the geopolitical thinking done in so-called think tanks.

In elaborate computer studies, it has emerged that one of the best strategies to pursue in all sorts of games with multiple rounds is called Tit-for-Tat. You pick the cooperative strategy that says: Cooperate, keep cooperating until you get double-crossed, then *you* clobber your opponent, and then if the opponent goes back to cooperating, you cooperate, too. This strategy beats selfish strategies, and It beats random strategies of cooperation versus betrayal. Remember, you read it here first: Be a nice guy as much as you can, but nuke 'em when necessary. It's good to know this, given that for all practical purposes you are stuck in dozens of ongoing multiple-round games every day.

Chaos

In the 18th and 19th centuries, the mathematicians who developed the study of differential equations were obliged to look for equations with solutions that could be pursued with paper and pencil. They did wonderful work solving *linear* versions of the equations that are important in physics, such as the wave equation. Linear in this context pretty much means "well behaved"—small changes in conditions make small changes in the results of the equations.

With the appearance of high-speed computers, mathematicians could begin investigating *nonlinear* differential equations, which often exhibit strange behavior. An American meteorologist named Edward Lorenz, to take just one example, decided to investigate a simple but nonlinear set of equations that model certain weather conditions. He found, to everyone's surprise, that tiny, barely measurable differences in the starting conditions for the weather system produced unrecognizably different results in the model after a few days. The equations of the model are said to be *chaotic,* unlike the equations in the well-behaved models of the past.

A shocking lot of mind-numbing nonsense has been written about chaos in the last decade. If you're interested at all, still the best starting point for novice readers is the book *Chaos* by James Gleick. Our new computer ability to model difficult equations comes with many serious implications, but in general the implications are nearly opposite to the often-heard popular-press generalizations that "chaos proves that most processes are random and unpredictable."

Fractals

Fractal curves are a popular topic in personal computing because a computer with decent color (256 colors at least) can make spectacular, almost hypnotic displays of two-dimensional fractal curves. Even in black and white, a complex fractal image is fascinating, as Figure 25-14 shows.

Figure 25-14:
A flashy
fractal.

Figure 25-15 shows a very simple fractal curve called the *Koch snowflake*. Start with a triangle and put little triangles on the faces of it. Then, on each triangle side, make more little triangles. Keep doing this until you're delirious.

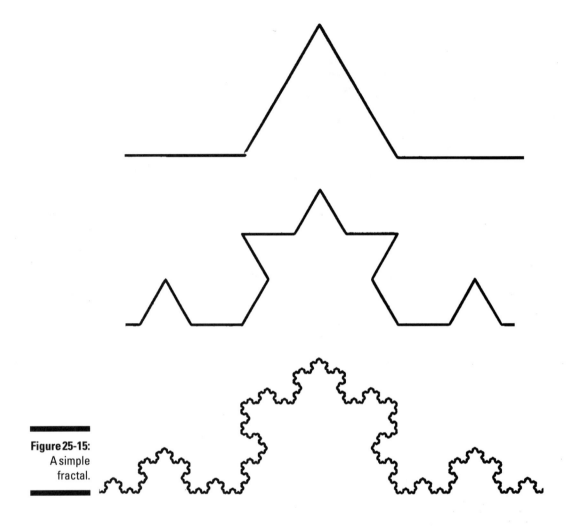

Figure 25-15:
A simple
fractal.

What do you think is the length of the line that constitutes the boundary of the snowflake? As you zoom in, you see that each little part of the snowflake is similar to each bigger part, because all parts are constructed along the same rule. Every round of repeating triangles makes the boundary somewhat longer, but does this boundary length approach a limit or just keep slowly increasing?

This question is answered, and your taste for cool fractal pictures should be satisfied, in *The Beauty of Fractals* by H. O. Peitgen and P. H. Richter (Springer, 1986). As a fun fact, many people believe that the mathematician in the Michael Crichton book *Jurassic Park* (played by Jeff Goldblum in the movie) was modeled on Herr Dr. Peitgen.

Fermat's Theorem

Pierre Fermat, a French jurist of the 17th century who made large contributions to mathematics, left as part of his legacy one particularly simple statement. We know, for example, that

$$3^2 + 4^2 = 5^2$$

and can find other examples as well where two squares add up to another square.

Fermat states the case

Fermat said that not only will there be no examples in whole numbers for

$$a^3 + b^3 = c^3$$

but there won't be any for higher powers, either. He wrote this observation in the margin of a text and then, in one of the truly great throw-away lines in math or any other endeavor, said he had discovered a marvelous proof of this statement but that the margin was too small to contain it.

Three hundred years later

You can only wish that the book had margins the size of a bed sheet, because 300 years later nobody had made much of a dent in proving this statement. Not that anyone doubted it — computer searches of numbers for exponents up to huge values had never turned up a counter-example to prove it wrong. But all the great mathematicians of three centuries failed to prove this remarkably simple theorem, and most of them figured that Fermat had discovered one of the many simple "proofs" that turn out not to be proofs when investigated more closely. If Gauss, Euler, Hilbert, Ramanujan, and von Neumann couldn't prove this statement, it's no insult to good old Fermat to conjecture that he didn't really have a proof either.

Finally, after one false start, in 1994 the English mathematician and Princeton professor Andrew Wiles produced a proof that has been verified by a number of other mathematicians. The proof is far from simple (in compact form, it runs about 200 pages) and has practically nothing to do with any other elementary proofs in the theory of numbers (that's the branch of mathematics containing this particular problem). None of humankind's outstanding material problems have been resolved by this proof, but as an intellectual exercise for its own sake, it's a real score for the human spirit.

Proofs and Computers

The proof of Fermat's "Last Theorem," as it's called, is one of the most impressive demonstrations in recent years of traditional mathematical proof. These days, quite a bit of mathematical research is conducted on objects that exist on computers, and often proofs have a large computer component.

Few people question that experimental mathematics on a computer has lots of valid uses. On a computer, for example, a mathematician can explore the properties of geometric objects in a seven-dimensional space, a realm in which most peoples' imaginations begin to fail them. Computer graphics are a great help in visualizing geometry, even in ordinary three-dimensional problems.

There are serious questions, however, about proofs in which most of the work is the result of a computer program. If the computer investigated all the cases necessary to prove a theorem, is it clear that any human really understands what's going on? How does one go about proving that the computer program used was itself entirely correct? Proving program correctness is a notoriously difficult problem, at least beyond the simplest programs.

None of this is a big concern in everyday math, where half the questions concern money and are answered with arithmetic that was already old when the pharaoh Tutankhamen was buried. But I bring it up here because you can often get the impression from school mathematics that math in general is just there; that it's finished and that not much new is happening. Nothing could be further from the truth. Mathematics is in a state of upheaval these days, with new topics and new controversies appearing every few months. A lot is happening at the frontiers of math, and it's not that far from the core topics to the frontiers.

Index

⚫ ⚫

(continued)

(continued)

(continued)

(continued)

Notes

Notes

Notes

Notes

Notes

Notes

Notes

Title	Author	ISBN	Price
INTERNET / COMMUNICATIONS / NETWORKING			12/20/94
CompuServe For Dummies™	by Wallace Wang	1-56884-181-7	$19.95 USA/$26.95 Canada
Modems For Dummies™, 2nd Edition	by Tina Rathbone	1-56884-223-6	$19.99 USA/$26.99 Canada
Modems For Dummies™	by Tina Rathbone	1-56884-001-2	$19.95 USA/$26.95 Canada
MORE Internet For Dummies™	by John R. Levine & Margaret Levine Young	1-56884-164-7	$19.95 USA/$26.95 Canada
NetWare For Dummies™	by Ed Tittel & Deni Connor	1-56884-003-9	$19.95 USA/$26.95 Canada
Networking For Dummies™	by Doug Lowe	1-56884-079-9	$19.95 USA/$26.95 Canada
ProComm Plus 2 For Windows For Dummies™	by Wallace Wang	1-56884-219-8	$19.99 USA/$26.99 Canada
The Internet For Dummies™, 2nd Edition	by John R. Levine & Carol Baroudi	1-56884-222-8	$19.99 USA/$26.99 Canada
The Internet For Macs For Dummies™	by Charles Seiter	1-56884-184-1	$19.95 USA/$26.95 Canada
MACINTOSH			
Macs For Dummies®	by David Pogue	1-56884-173-6	$19.95 USA/$26.95 Canada
Macintosh System 7.5 For Dummies™	by Bob LeVitus	1-56884-197-3	$19.95 USA/$26.95 Canada
MORE Macs For Dummies™	by David Pogue	1-56884-087-X	$19.95 USA/$26.95 Canada
PageMaker 5 For Macs For Dummies™	by Galen Gruman	1-56884-178-7	$19.95 USA/$26.95 Canada
QuarkXPress 3.3 For Dummies™	by Galen Gruman & Barbara Assadi	1-56884-217-1	$19.99 USA/$26.99 Canada
Upgrading and Fixing Macs For Dummies™	by Kearney Rietmann & Frank Higgins	1-56884-189-2	$19.95 USA/$26.95 Canada
MULTIMEDIA			
Multimedia & CD-ROMs For Dummies™, Interactive Multimedia Value Pack	by Andy Rathbone	1-56884-225-2	$29.95 USA/$39.95 Canada
Multimedia & CD-ROMs For Dummies™	by Andy Rathbone	1-56884-089-6	$19.95 USA/$26.95 Canada
OPERATING SYSTEMS / DOS			
MORE DOS For Dummies™	by Dan Gookin	1-56884-046-2	$19.95 USA/$26.95 Canada
S.O.S. For DOS™	by Katherine Murray	1-56884-043-8	$12.95 USA/$16.95 Canada
OS/2 For Dummies™	by Andy Rathbone	1-878058-76-2	$19.95 USA/$26.95 Canada
UNIX			
UNIX For Dummies™	by John R. Levine & Margaret Levine Young	1-878058-58-4	$19.95 USA/$26.95 Canada
WINDOWS			
S.O.S. For Windows™	by Katherine Murray	1-56884-045-4	$12.95 USA/$16.95 Canada
MORE Windows 3.1 For Dummies™, 3rd Edition	by Andy Rathbone	1-56884-240-6	$19.99 USA/$26.99 Canada
PCs / HARDWARE			
Illustrated Computer Dictionary For Dummies™	by Dan Gookin, Wally Wang, & Chris Van Buren	1-56884-004-7	$12.95 USA/$16.95 Canada
Upgrading and Fixing PCs For Dummies™	by Andy Rathbone	1-56884-002-0	$19.95 USA/$26.95 Canada
PRESENTATION / AUTOCAD			
AutoCAD For Dummies™	by Bud Smith	1-56884-191-4	$19.95 USA/$26.95 Canada
PowerPoint 4 For Windows For Dummies™	by Doug Lowe	1-56884-161-2	$16.95 USA/$22.95 Canada
PROGRAMMING			
Borland C++ For Dummies™	by Michael Hyman	1-56884-162-0	$19.95 USA/$26.95 Canada
"Borland's New Language Product" For Dummies™	by Neil Rubenking	1-56884-200-7	$19.95 USA/$26.95 Canada
C For Dummies™	by Dan Gookin	1-878058-78-9	$19.95 USA/$26.95 Canada
C++ For Dummies™	by Stephen R. Davis	1-56884-163-9	$19.95 USA/$26.95 Canada
Mac Programming For Dummies™	by Dan Parks Sydow	1-56884-173-6	$19.95 USA/$26.95 Canada
QBasic Programming For Dummies™	by Douglas Hergert	1-56884-093-4	$19.95 USA/$26.95 Canada
Visual Basic "X" For Dummies™, 2nd Edition	by Wallace Wang	1-56884-230-9	$19.99 USA/$26.99 Canada
Visual Basic 3 For Dummies™	by Wallace Wang	1-56884-076-4	$19.95 USA/$26.95 Canada
SPREADSHEET			
1-2-3 For Dummies™	by Greg Harvey	1-878058-60-6	$16.95 USA/$21.95 Canada
1-2-3 For Windows 5 For Dummies™, 2nd Edition	by John Walkenbach	1-56884-216-3	$16.95 USA/$21.95 Canada
1-2-3 For Windows For Dummies™	by John Walkenbach	1-56884-052-7	$16.95 USA/$21.95 Canada
Excel 5 For Macs For Dummies™	by Greg Harvey	1-56884-186-8	$19.95 USA/$26.95 Canada
Excel For Dummies™, 2nd Edition	by Greg Harvey	1-56884-050-0	$16.95 USA/$21.95 Canada
MORE Excel 5 For Windows For Dummies™	by Greg Harvey	1-56884-207-4	$19.95 USA/$26.95 Canada
Quattro Pro 6 For Windows For Dummies™	by John Walkenbach	1-56884-174-4	$19.95 USA/$26.95 Canada
Quattro Pro For DOS For Dummies™	by John Walkenbach	1-56884-023-3	$16.95 USA/$21.95 Canada
UTILITIES / VCRs & CAMCORDERS			
Norton Utilities 8 For Dummies™	by Beth Slick	1-56884-166-3	$19.95 USA/$26.95 Canada
VCRs & Camcorders For Dummies™	by Andy Rathbone & Gordon McComb	1-56884-229-5	$14.99 USA/$20.99 Canada
WORD PROCESSING			
Ami Pro For Dummies™	by Jim Meade	1-56884-049-7	$19.95 USA/$26.95 Canada
MORE Word For Windows 6 For Dummies™	by Doug Lowe	1-56884-165-5	$19.95 USA/$26.95 Canada
MORE WordPerfect 6 For Windows For Dummies™	by Margaret Levine Young & David C. Kay	1-56884-206-6	$19.95 USA/$26.95 Canada
MORE WordPerfect 6 For DOS For Dummies™	by Wallace Wang, edited by Dan Gookin	1-56884-047-0	$19.95 USA/$26.95 Canada
S.O.S. For WordPerfect™	by Katherine Murray	1-56884-053-5	$12.95 USA/$16.95 Canada
Word 6 For Macs For Dummies™	by Dan Gookin	1-56884-190-6	$19.95 USA/$26.95 Canada
Word For Windows 6 For Dummies™	by Dan Gookin	1-56884-075-6	$16.95 USA/$21.95 Canada
Word For Windows For Dummies™	by Dan Gookin	1-878058-86-X	$16.95 USA/$21.95 Canada
WordPerfect 6 For Dummies™	by Dan Gookin	1-878058-77-0	$16.95 USA/$21.95 Canada
WordPerfect For Dummies™	by Dan Gookin	1-878058-52-5	$16.95 USA/$21.95 Canada
WordPerfect For Windows For Dummies™	by Margaret Levine Young & David C. Kay	1-56884-032-2	$16.95 USA/$21.95 Canada

Fun, Fast, & Cheap!

CorelDRAW! 5 For Dummies™ Quick Reference
by Raymond E. Werner

ISBN: 1-56884-952-4
$9.99 USA/$12.99 Canada

Windows "X" For Dummies™ Quick Reference, 3rd Edition
by Greg Harvey

ISBN: 1-56884-964-8
$9.99 USA/$12.99 Canada

Word For Windows 6 For Dummies™ Quick Reference
by George Lynch

ISBN: 1-56884-095-0
$8.95 USA/$12.95 Canada

WordPerfect For DOS For Dummies™ Quick Reference
by Greg Harvey

ISBN: 1-56884-009-8
$8.95 USA/$11.95 Canada

Title	Author	ISBN	Price
DATABASE			
Access 2 For Dummies™ Quick Reference	by Stuart A. Stuple	1-56884-167-1	$8.95 USA/$11.95 Canada
dBASE 5 For DOS For Dummies™ Quick Reference	by Barry Sosinsky	1-56884-954-0	$9.99 USA/$12.99 Canada
dBASE 5 For Windows For Dummies™ Quick Reference	by Stuart J. Stuple	1-56884-953-2	$9.99 USA/$12.99 Canada
Paradox 5 For Windows For Dummies™ Quick Reference	by Scott Palmer	1-56884-960-5	$9.99 USA/$12.99 Canada
DESKTOP PUBLISHING / ILLUSTRATION/GRAPHICS			
Harvard Graphics 3 For Windows For Dummies™ Quick Reference	by Raymond E. Werner	1-56884-962-1	$9.99 USA/$12.99 Canada
FINANCE / PERSONAL FINANCE			
Quicken 4 For Windows For Dummies™ Quick Reference	by Stephen L. Nelson	1-56884-950-8	$9.95 USA/$12.95 Canada
GROUPWARE / INTEGRATED			
Microsoft Office 4 For Windows For Dummies™ Quick Reference	by Doug Lowe	1-56884-958-3	$9.99 USA/$12.99 Canada
Microsoft Works For Windows 3 For Dummies™ Quick Reference	by Michael Partington	1-56884-959-1	$9.99 USA/$12.99 Canada
INTERNET / COMMUNICATIONS / NETWORKING			
The Internet For Dummies™ Quick Reference	by John R. Levine	1-56884-168-X	$8.95 USA/$11.95 Canada
MACINTOSH			
Macintosh System 7.5 For Dummies™ Quick Reference	by Stuart J. Stuple	1-56884-956-7	$9.99 USA/$12.99 Canada
OPERATING SYSTEMS / DOS			
DOS For Dummies® Quick Reference	by Greg Harvey	1-56884-007-1	$8.95 USA/$11.95 Canada
UNIX			
UNIX For Dummies™ Quick Reference	by Margaret Levine Young & John R. Levine	1-56884-094-2	$8.95 USA/$11.95 Canada
WINDOWS			
Windows 3.1 For Dummies™ Quick Reference, 2nd Edition	by Greg Harvey	1-56884-951-6	$8.95 USA/$11.95 Canada
PRESENTATION / AUTOCAD			
AutoCAD For Dummies™ Quick Reference	by Ellen Finkelstein	1-56884-198-1	$9.95 USA/$12.95 Canada
SPREADSHEET			
1-2-3 For Dummies™ Quick Reference	by John Walkenbach	1-56884-027-6	$8.95 USA/$11.95 Canada
1-2-3 For Windows 5 For Dummies™ Quick Reference	by John Walkenbach	1-56884-957-5	$9.95 USA/$12.95 Canada
Excel For Windows For Dummies™ Quick Reference, 2nd Edition	by John Walkenbach	1-56884-096-9	$8.95 USA/$11.95 Canada
Quattro Pro 6 For Windows For Dummies™ Quick Reference	by Stuart A. Stuple	1-56884-172-8	$9.95 USA/$12.95 Canada
WORD PROCESSING			
Word For Windows 6 For Dummies™ Quick Reference	by George Lynch	1-56884-095-0	$8.95 USA/$11.95 Canada
WordPerfect For Windows For Dummies™ Quick Reference	by Greg Harvey	1-56884-039-X	$8.95 USA/$11.95 Canada

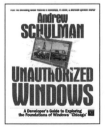
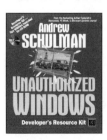

ORDER FORM

IDG BOOKS

Order Center: **(800) 762-2974** *(8 a.m.–6 p.m., EST, weekdays)*

Quantity	ISBN	Title	Price	Total

Shipping & Handling Charges

	Description	First book	Each additional book	Total
Domestic	Normal	$4.50	$1.50	$
	Two Day Air	$8.50	$2.50	$
	Overnight	$18.00	$3.00	$
International	Surface	$8.00	$8.00	$
	Airmail	$16.00	$16.00	$
	DHL Air	$17.00	$17.00	$

*For large quantities call for shipping & handling charges.
**Prices are subject to change without notice.

Ship to:

Name _____

Company _____

Address _____

City/State/Zip _____

Daytime Phone _____

Payment: n Check to IDG Books (US Funds Only)
n VISAn MasterCard n American Express

Card # _____ Expires _____

Signature _____

Subtotal _____

CA residents add
applicable sales tax _____

IN, MA, and MD
residents add
5% sales tax _____

IL residents add
6.25% sales tax _____

RI residents add
7% sales tax _____

TX residents add
8.25% sales tax _____

Shipping _____

Total _____

Please send this order form to:

IDG Books Worldwide
7260 Shadeland Station, Suite 100
Indianapolis, IN 46256

Allow up to 3 weeks for delivery.
Thank you!